CREATIVE EVOLUTION

Creative Evolution

By *HENRI BERGSON*

In the Authorized Translation

by ARTHUR MITCHELL

With a Foreword by IRWIN EDMAN

GREENWOOD PRESS, PUBLISHERS
WESTPORT, CONNECTICUT

Library of Congress Cataloging in Publication Data

Bergson, Henri Louis, 1859-1941.
Creative evolution.

Reprint of the 1944 ed. published by Modern Library, New York, in series: The Modern library of the world's best books.
Includes index.
1. Metaphysics. 2. Life. 3. Evolution. I. Title.
B2430.B4E72 1975 113 74-28524
ISBN 0-8371-7917-3

This edition originally published in 1944 by Random House, Inc.

This edition is reprinted by arrangement with Holt, Rinehart and Winston, Inc.

Reprinted by Greenwood Press, Inc.

First Greenwood reprinting 1975
Second Greenwood reprinting 1977

Library of Congress catalog card number 74-28524
ISBN 0-8371-7917-3

Printed in the United States of America

TRANSLATOR'S NOTE

In the writing of this English translation of Professor Bergson's most important work, I was helped by the friendly interest of Professor William James, to whom I owe the illumination of much that was dark to me as well as the happy rendering of certain words and phrases for which an English equivalent was difficult to find. His sympathetic appreciation of Professor Bergson's thought is well known, and he has expressed his admiration for it in one of the chapters of *A Pluralistic Universe*. It was his intention, had he lived to see the completion of this translation, himself to introduce it to English readers in a prefatory note.

I wish to thank my friend, Dr. George Clarke Cox, for many valuable suggestions.

I have endeavored to follow the text as closely as possible, and at the same time to preserve the living union of diction and thought. Professor Bergson has himself carefully revised the whole work. We both of us wish to acknowledge the great assistance of Miss Millicent Murby. She has kindly studied the translation phrase by phrase, weighing each word, and her revision has resulted in many improvements.

But above all we must express our acknowledgment to Mr. H. Wildon Carr, the Honorary Secretary of the Aristotelian Society of London, and the writer of several studies of "Evolution Creatrice."[1] We asked him to be kind enough to revise the proofs of our work. He has done much more than revise them: they have come from his hands with his personal mark in many places. We cannot express all that the present work owes to him.

ARTHUR MITCHELL

HARVARD UNIVERSITY

[1] *Proceedings of the Aristotelian Society*, vols. ix. and x., and *Hibbert Journal* for July, 1910.

CONTENTS

CHAPTER I

The Evolution of Life—Mechanism and Teleology

CHAPTER II

The Divergent Directions of the Evolution of Life—Torpor, Intelligence, Instinct

CHAPTER III

On the Meaning of Life—The Order of Nature and the Form of Intelligence

CHAPTER IV

The Cinematographical Mechanism of Thought and the Mechanistic Illusion—A Glance at the History of Systems—Real Becoming and False Evolutionism

FOREWORD

by Irwin Edman

HENRI BERGSON died in Paris on January 4, 1941, at the age of eighty-one. The general public, and that relatively private clique known as the philosophical public, had long ago fallen into the habit of thinking of Bergson as dead. Only on the publication of the dramatic news of Bergson's decision to renounce all posts and honors rather than to accept exemption from the anti-Semitic laws of the Vichy government was the world reminded that he was still alive. There had been an interval of more than twenty-five years between the publication of his phenomenally successful *Creative Evolution* in 1907 and the book rather blindly called *The Two Sources of Morality and Religion*. A few people knew vaguely that Bergson was still living, an invalid in retirement, in Paris. A few, and not only philosophers, remembered when his name was something to conjure with, and his philosophy was hailed, by William James among others, as an almost medicinal magic. "Oh, my Bergson," James wrote to the author of *Creative Evolution* when that book first appeared, "you are a magician and your book is a marvel, a real wonder. . . . But, unlike the works of genius of the Transcendentalist movement (which are so obscurely and abominably and inaccessibly written), a pure classic in point of form

. . . such a flavor of persistent euphony, as of a rich river that never foamed or ran thin, but steadily and firmly proceeded with its banks full to the brim. Then the aptness of your illustrations, that never scratch or stand out at right angles, but invariably simplify the thought and help to pour it along. Oh, indeed you are a magician! And if your next book proves to be as great an advance on this one as this is on its two predecessors, your name will surely go down as one of the great creative names in philosophy."

James's generous salute came at the very peak of Bergson's reputation. There was a whole epidemic of books about Bergson; his themes and his terms became clichés in philosophical discussion all over the world; and all over the world, too, *Creative Evolution,* with its gleaming mellifluous stream of thought, entranced many more than could understand it, and many readers, too, of many different worlds: the fashionable dowagers who found refuge from boredom in his *élan vital,* the religious liberals welcoming a philosopher who seemed to have found critical circumvention of mechanistic science and a new and poetic support for belief in God, in free will, and even, though in a somewhat Pickwickian sense, in immortality. Sorel could connect Bergson's theory of reality as an integral movement with a doctrine of revolution as contrasted with piecemeal parliamentary reform. Those wearied of the rationalism of the Transcendentalists and the fixities and the iron necessities of materialistic science found in him a hope, an inspiration and a release. Here, moreover, was one romantic who seemed to have a clear head. Here, too, was a philosopher who, while a consummate literary artist, could beat precise analysts at their own game, and analyze nuances of feeling, action and thought which seemed to escape through the nets of logic and to be crushed by the tight schemes of matter in motion.

All that seems long ago now. Bergson, secure, or so it seemed in the years before the Vichy government was thought of, in his post at the Collège de France, had his claim to immortality staked out. But his popularity with the general literate public vanished almost as quickly as it had appeared. As for the professional philosophers whose admiration had always been tinctured with critical reserve, these now found more to criticize and, what was perhaps more decisive, began to think in other terms about other problems. When, in 1935, the English edition of *Les Deux Sources de la Religion et de la Morale* appeared, having been completed through long and painful years of paralyzing illness, all the masteries and subtleties of analysis were still there, extended now to morals and religion, and to art as well. The book was respectfully greeted, and has had some academic influence, but Bergson's day as a central figure in philosophy and, certainly, in general culture seemed to be over. Like other intellectual figures who became briefly fashionable, he had become dated before he died. All that remained in the public memory were a few tags: "*élan vital*," "creative evolution," "the stream of consciousness," "the flow of reality." Now other men were using other terms for other issues. The intellectual scene and the world setting in which Bergson wrote had almost nightmarishly changed.

What did Bergson contribute? What remains? Looking back now, one sees that, for all his "French" clarity

(Bergson is one of the most lucid writers ever to have been a professor of philosophy), Bergson is in a romantic, almost a German romantic, tradition. In his early youth he was a disciple of Herbert Spencer, but soon felt, as so many less articulate people in the late nineteenth century felt, something sterile and inadequate, something artificial about the "reality" revealed by "scientific laws." In his youth Bergson had studied psychology and biology, and was caught by the fascination of two ideas, nay, for him two realities: life and time. One of the most characteristically contagious passages of his works is in his little *Introduction to Metaphysics*, where he makes one feel and realize, almost as a poet might, the tension and the fluency of time, the urgency and poignancy of duration. Bergson was rebelling against the fixities and rigidities which both logicians and materialists had ascribed to reality. Bergson found reality in movement and change themselves, an *aperçu* not uncongenial to the dynamic changing society in which he lived. If change was real, novelty was real; if novelty was real, freedom was real. The immediate was flux, and the changing was ultimate. When in *Creative Evolution* Bergson turned to biological considerations, he held that change means growth, growth means creation, creation means freedom. And if freedom was ultimately real, what a liberation that spelled for the soul of man, no longer bound by the fixities of space, of logic and of habit! The real facts of evolution were to be found, not in a mechanical elimination of the unfit, but in the creative surge of life, in an *élan vital*. That propulsive life was best known in the living of it, "bathing in the full stream of experience." Knowledge was not in spatial formulas, post-mortems of the living flux; knowledge lay in intuition, in the self-immolating and absorbed insights of the poet, the artist, the saint, of men at the acme of life, in the creative activity of genius, worship or love.

To the professional philosopher, Bergson's most acute and permanent contributions came not from his celebrations of the aspiring and inventive movement of evolution or the novel cumulative growth of nature. They lay, rather, in those early works, *Introduction to Metaphysics*, and *Matter and Memory*, wherein Bergson developed his metaphysics and epistemology of change and indetermination. Actuality, reality, lay at the junctions of experience, the acutely felt centers of indetermination which marked those successive births and rebirths of awareness we call consciousness. Awareness itself is a function of the possibility of choice; it is, in fact, the sense of choice. As experience cumulates, any moment focuses wider and wider, deeper and deeper, more richly concentrated resources of choice and felt possibilities of action. Memory in action is not a dead deposit; it is a living and functional focusing of energies. It is life at the acme of attention, creation and decision. Memory is life cumulated and brought to bear as alternatives of action, as impellingly realized possibilities of choice. Memory is the living reality, the past felt, those moments of heightened consciousness which we feel as suggested opportunities to make the future. What we commonly call memory is what we commonly —and falsely—call time: spatially measurable images.

Matter is the deposit of life, the static residues of actions done, choices made in the past. Living memory is the past felt in the actualities of realities, of change.

In *Time and Free Will* Bergson elaborates with skillful dialectic detail—Bergson's mastery of his technical tools is one of his most impressive virtuoso characteristics—time, "duration," in contrast with measured and the measurable "physical" time of the physicists. He traverses here the same theme, in essence, with which he is concerned in *Matter and Memory*. Here time as real is again time as experienced as freedom, as, and in, moments of choice. Necessity is the order of the fixed, of the past as analyzed by the retrospective, geometrizing intellect.

Bergson's theory of time and memory becomes a critique of purposes and ends in the logic and metaphysics of both idealism and mechanism, and of both mechanical and vitalistic evolution. Evolution moves to no fixed ends, and, on the other hand, it is more than advance by mechanical Darwinian elimination. Evolution (and all of nature) is the expression of a creative urgency, a fulfillment of novel ends inventively generated in the process of time itself. Bergson's critique of the purely intellectual analysis of the idealists and the purely geometrical analysis of the materialists is shrewd criticism, however vague the positive doctrine of "creative evolution."

Not least suggestive is Bergson's brilliant critique of pure logic and pure mechanism in morals, art and religion. He naturalizes the saint, the artist and the prophet. These stand at junctures of openness and freshness. Necessity, obligation and formalism close minds and close societies. Genius and vision open new roads and new perspectives. Prophets are the creative memories of the race. Bergson at the end remained the voice, both romantic and critical, of the creative impulses in societies given over to regimentation and formulas. For Bergson, himself a lucid intellectual, was on the side of the visionary poets, seers and saints.

It is not hard to see what the dreaming spirits of men, whose dreams had been clipped by physics and by society, found in this celebration of enraptured impulse and creative movement. It is easy to see what people, wearied of fixed and conventional goals in what Bergson in *Morality and Religion* was to call a "closed society," found to cherish in his conception of spontaneous freedom in open societies. In the latter, Bergson suggested, the intuitions of seers and saints and poets opened new roads and suggested unattained but not impossible heights from which men might have angelic vistas. "The universe," he says at the close of *Morality and Religion,* ". . . is a machine for the making of God."

Bergson's persuasiveness came not so much from the seductive vagueness of intuition and of the *élan vital* as from his shrewd and subtly destructive analyses of the pretensions of the intellect and of intellectualism to be revelations of reality. Intelligence—here he was at one with the pragmatists—was purely practical. It set up immobilities to guide us in the chartless flux. But for the truth we must turn our backs on the "false secondary power by which we multiply distinctions," and feel intimately the pulsing movement of life itself. The pragmatists enjoyed and applauded Bergson's critique of intelligence and scientific method. The mystics and the romanticists applauded and hugged to their hearts the passionate hugging of life which he counseled as the way of reality. The liberal theologians were delighted because they thought the *élan vital* might lead anywhere. Did not Bergson himself suggest that it might lead to immortality and to God? In an age given to statistics and formulas the artists welcomed intuition.

Everyone rebelling against convention in conduct, chafing against formalism in art, revolting against the fixed and stable in thought, found in him an enchanting voice. Only other philosophers found things to grumble at. They did not mind Bergson's celebrating life, but they did not approve the obscurantism of turning one's back on intelligent discrimination. They admired the subtlety of his analyses, but could not make out where the analysis led, or for that matter, where the *élan vital* led. They questioned the casuistry by which Bergson proved there was a real future without a causal past. They distrusted, above all, the anti-intellectualism to which Bergson turned his own fine intellect. The *élan vital* means a renaissance to a poet; to a barbarian it means brute power. The reactionary forces now in control of France are also exhibitions of the *élan vital*. But it was the intelligence of integrity that caused Bergson, just a few weeks before his death, to refuse to be made an exception by the Vichy government to their racial laws. Bergson's philosophy was once hailed as a new thing in the world. Its elements are very old; its mysticism is as old as Plotinus, to whom Bergson acknowledged himself much indebted. His *élan vital* goes back a long way too: ultimately to the Dionysiac mysteries and, in the modern world, to Schopenhauer. Its romanticism goes back to Schelling, Fichte and Rousseau. Its dress was sophisticated and new; its substance was old and primitive, in both vitality and opaqueness.

There are subterranean vital forces now come to the surface of the world. The world is a machine making devils as well as gods. The abdication of mind is celebrated by barbarians. Instinct and intuition demand education, not wanton trust. Analysis may be abstract or irrelevant, as Bergson pointed out; it may also be the servant of that life which otherwise wanders into dark by-paths and fanatic blind alleys. "Evolution" is creative when men intelligently co-operate; otherwise it is, as events have proved, brutal, fatal and blind.

One might have selected another book of Bergson's for inclusion in the Modern Library. *Time and Free Will,* for example, is in some ways a subtler piece of analysis and a far more permanent contribution to the study of psychological aspects of experience than Bergson's *Creative Evolution* is a contribution to the understanding of its biological phases. But nearly everything Bergson has to say about other things, even about religion, politics and art, is implicit in these pages. It contains a very brilliant critique of the assumptions of mechanistic science; it conveys the intimate sense of reality as life, of life as movement, of movement as creation, which is one of Bergson's central themes. It is marked by his extraordinary gift for philosophical images, such as that of life being the skyrocket bursting in air, and matter being the dead ashes falling down, or again of reality being a line in the drawing rather than the line drawn. It contains his apotheosis of man as the tool-making animal and his treatment of intelligence as itself a tool which helps man in his practical necessities while it hides from him the moving reality of being which the path of instinct and intuition, the alternative to intelligence, discloses. It suggests all he later says explicitly about the creative functions of the artist, the saint and the authentic revolutionary, awake to, inventive of, the fresh, the alive breaking through the encrustations of government, society and learning.

In perspective now Bergson, as recent scholarship has shown, is less daringly original than William James and others had supposed. But the note of creation and its spirit, which is struck in these pages, is in all Bergson's books. *Creative Evolution* by its very title and by the hold it has exercised over the educated public has about it the atmosphere, not illusory, of a classic, a modern one, and, one suspects, a permanent one. Every author has his note which comes out most clearly and comprehensively in one of his books. Bergson's note and most of his overtones are in *Creative Evolution.*

December, 1943

INTRODUCTION

THE history of the evolution of life, incomplete as it yet is, already reveals to us how the intellect has been formed, by an uninterrupted progress, along a line which ascends through the vertebrate series up to man. It shows us in the faculty of understanding an appendage of the faculty of acting, a more and more precise, more and more complex and supple adaptation of the consciousness of living beings to the conditions of existence that are made for them. Hence should result this consequence that our intellect, in the narrow sense of the word, is intended to secure the perfect fitting of our body to its environment, to represent the relations of external things among themselves—in short, to think matter. Such will indeed be one of the conclusions of the present essay. We shall see that the human intellect feels at home among inanimate objects, more especially among solids, where our action finds its fulcrum and our industry its tools; that our concepts have been formed on the model of solids; that our logic is, pre-eminently, the logic of solids; that, consequently, our intellect triumphs in geometry, wherein is revealed the kinship of logical thought with unorganized matter, and where the intellect has only to follow its natural movement, after the lightest possible contact with experience, in order to go from discovery to discovery, sure that experience is following behind it and will justify it invariably.

But from this it must also follow that our thought, in its purely logical form, is incapable of presenting the true nature of life, the full meaning of the evolutionary movement. Created by life, in definite circumstances, to act on definite things, how can it embrace life, of which it is only an emanation or an aspect? Deposited by the evolutionary movement in the course of its way, how can it be applied to the evolutionary movement itself? As well contend that the part is equal to the whole, that the effect can reabsorb its cause, or that the pebble left on the beach displays the form of the wave that brought it there. In fact, we do indeed feel that not one of the categories of our thought—unity, multiplicity, mechanical causality, intelligent finality, etc.—applies exactly to the things of life: who can say where individuality begins and ends, whether the living being is one or many, whether it is the cells which associate themselves into the organism or the organism which dissociates itself into cells? In vain we force the living into this or that one of our molds. All the molds crack. They are too narrow, above all too rigid, for what we try to put into them. Our reasoning, so sure of itself among things inert, feels ill at ease on this new ground. It would be difficult to cite a biological discovery due to pure reasoning. And most often, when experience has finally shown us how life goes to work to obtain a certain result, we find its way of working is just that of which we should never have thought.

Yet evolutionist philosophy does not hesitate to extend to the things of life the same methods of explanation which have succeeded in the case of unorganized matter. It begins by showing us in the intellect a local effect of evolution, a flame, perhaps accidental, which lights up the coming and going of living beings in the narrow passage open to their action; and lo! forgetting what it has just told us, it makes of this lantern glimmering in a tunnel a Sun which can illuminate the world. Boldly it proceeds, with the powers of conceptual thought alone, to the ideal reconstruction of all things, even of life. True, it hurtles in its course against such formidable difficulties, it sees its logic end in such strange contradictions, that it very speedily renounces its first ambition. "It is no longer reality itself," it says, "that it will reconstruct, but only an imitation of the real, or rather a symbolical image; the essence of things escapes us, and will escape us always; we move among relations; the absolute is not in our province; we are brought to a stand before the Unknowable."—But for the human intellect, after too much pride, this is really an excess of humility. If the intellectual form of the living being has been gradually modeled on the reciprocal actions and reactions of certain bodies and their material environment, how should it not reveal to us something of the very essence of which these bodies are made? Action cannot move in the unreal. A mind born to speculate or to dream, I admit, might remain outside reality, might deform or transform the real, perhaps even create it—as we create the figures of men and animals that our imagination cuts out of the passing cloud. But an intellect bent upon the act to be performed and the reaction to follow, feeling its object so as to get its mobile impression at every instant, is an intellect that touches something of the absolute. Would the idea ever have occurred to us to doubt this absolute value of our knowledge if philosophy had not shown us what contradictions our speculation meets, what dead-locks it ends in? But these difficulties and contradictions all arise from trying to apply the usual forms of our thought to objects with which our industry has nothing to do, and for which, therefore, our molds are not made. Intellectual knowledge, in so far as it relates to a certain aspect of inert matter, ought, on the contrary, to give us a faithful imprint of it, having been stereotyped on this particular object. It becomes relative only if it claims, such as it is, to present to us life—that is to say, the maker of the stereotype-plate.

Must we then give up fathoming the depths of life? Must we keep to that mechanistic idea of it which the understanding will always give us—an idea necessarily artificial and symbolical, since it makes the total activity of life shrink to the form of a certain human activity which is only a partial and local manifestation of life, a result or by-product of the vital process? We should have to do so, indeed, if life had employed all the psychical potentialities it possesses in producing pure understandings—that is to say, in making geometricians. But the line of evolution that ends in man is not the only one. On other paths, divergent from it, other forms of consciousness have been developed, which have not been able to free themselves from external constraints or to regain control over themselves, as the human intellect has done, but which, none the less, also express something that is immanent and essential in the evolutionary movement. Suppose these other forms of consciousness brought together and amalgamated with intellect: would not the result be a consciousness as wide as life? And such a consciousness, turning around suddenly against the push of life which it feels behind, would have a vision of life complete—would it not?—even though the vision were fleeting.

It will be said that, even so, we do not transcend our intellect, for it is still with our intellect, and through our intellect, that we see the other forms of consciousness. And this would be right if we were pure intellects, if there did not remain, around our conceptual and logical thought, a vague nebulosity, made of the very substance out of which has been formed the luminous nucleus that we call the intellect. Therein reside certain powers that are complementary to the understanding, powers of which we have only an indistinct feeling when we remain shut up in ourselves, but which will become clear and distinct when they perceive themselves at work, so to speak, in the evolution of nature. They will thus learn what sort of effort they must make to be intensified and expanded in the very direction of life.

This amounts to saying that *theory of knowledge* and *theory of life* seem to us inseparable. A theory of life that is not accompanied by a criticism of knowledge is obliged to accept, as they stand, the concepts which the understanding puts at its disposal: it can but enclose the facts, willing or not, in pre-existing frames which it regards as ultimate. It thus obtains a symbolism which is convenient, perhaps even necessary to positive science, but not a direct vision of its object. On the other hand, a theory of knowledge which does not replace the intellect in the general evolution of life will teach us neither how the frames of knowledge have been constructed nor how we can enlarge or go beyond them. It is necessary that these two inquiries, theory of knowledge and theory of life, should join each other, and, by a circular process, push each other on unceasingly.

Together, they may solve by a method more sure, brought nearer to experience, the great problems that philosophy poses. For, if they should succeed in their common enterprise, they would show us the formation of the intellect, and thereby the genesis of that matter of which our intellect traces the general configuration. They would dig to the very root of nature and of mind. They would substitute for the false evolutionism of Spencer—which consists in cutting up present reality, already evolved, into little bits no less evolved, and then recomposing it with these fragments, thus positing in advance everything that is to be explained—a true evolutionism, in which reality would be followed in its generation and its growth.

But a philosophy of this kind will not be made in a day. Unlike the philosophical systems properly so called, each of which was the individual work of a man of genius and sprang up as a whole, to be taken or left, it will only be built up by the collective and progressive effort of many thinkers, of many observers also, completing, correcting and improving one another. So the present essay does not aim at resolving at once the greatest problems. It simply desires to define the method and to permit a glimpse, on some essential points, of the possibility of its application.

Its plan is traced by the subject itself. In the first

chapter, we try on the evolutionary progress the two ready-made garments that our understanding puts at our disposal, mechanism and finality;[1] we show that they do not fit, neither the one nor the other, but that one of them might be recut and resewn, and in this new form fit less badly than the other. In order to transcend the point of view of the understanding, we try, in our second chapter, to reconstruct the main lines of evolution along which life has traveled by the side of that which has led to the human intellect. The intellect is thus brought back to its generating cause, which we then have to grasp in itself and follow in its movement. It is an effort of this kind that we attempt—incompletely indeed—in our third chapter. A fourth and last part is meant to show how our understanding itself, by submitting to a certain discipline, might prepare a philosophy which transcends it. For that, a glance over the history of systems became necessary, together with an analysis of the two great illusions to which, as soon as it speculates on reality in general, the human understanding is exposed.

The idea of regarding life as transcending teleology as well as mechanism is far from being a new idea. Notably in three articles by Ch. Dunan on "Le problème de la vie" (*Revue philosophique*, 1892) it is profoundly treated. In the development of this idea, we agree with Ch. Dunan on more than one point. But the views we are presenting on this matter, as on the questions attaching to it, are those that we expressed long ago in our *Essai sur les données immédiates de la conscience* (Paris, 1889). One of the principal objects of that essay was, in fact, to show that the psychical life is neither unity nor multiplicity, that it transcends both the *mechanical* and the *intellectual*, mechanism and finalism having meaning only where there is "distinct multiplicity," "spatiality," and consequently assemblage of pre-existing parts: "real duration" signifies both undivided continuity and creation. In the present work we apply these same ideas to life in general, regarded, more over, itself from the psychological point of view.

CREATIVE EVOLUTION

CHAPTER I

THE EVOLUTION OF LIFE—MECHANISM AND TELEOLOGY

THE existence of which we are most assured and which we know best is unquestionably our own, for of every other object we have notions which may be considered external and superficial, whereas, of ourselves, our perception is internal and profound. What, then, do we find? In this privileged case, what is the precise meaning of the word "exist"? Let us recall here briefly the conclusions of an earlier work.

I find, first of all, that I pass from state to state. I am warm or cold, I am merry or sad, I work or I do nothing, I look at what is around me or I think of something else. Sensations, feelings, volitions, ideas—such are the changes into which my existence is divided and which color it in turns. I change, then, without ceasing. But this is not saying enough. Change is far more radical than we are at first inclined to suppose.

For I speak of each of my states as if it formed a block and were a separate whole. I say indeed that I change, but the change seems to me to reside in the passage from one state to the next: of each state, taken separately, I am apt to think that it remains the same during all the time that it prevails. Nevertheless, a slight effort of attention would reveal to me that there is no feeling, no idea, no volition which is not undergoing change every moment: if a mental state ceased to vary, its duration would cease to flow. Let us take the most stable of internal states, the visual perception of a motionless external object. The object may remain the same, I may look at it from the same side, at the same angle, in the same light; nevertheless the vision I now have of it differs from that which I have just had, even if only because the one is an instant older than the other. My memory is there, which conveys something of the past into the present. My mental state, as it advances on the road of time, is continually swelling with the duration which it accumulates: it goes on increasing—rolling upon itself, as a snowball on the snow. Still more is this the case with states more deeply internal, such as sensations, feelings, desires, etc., which do not correspond, like a simple visual perception, to an unvarying external object. But it is expedient to disregard this uninterrupted change, and to notice it only when it becomes sufficient to impress a new attitude on the body, a new direction on the attention. Then, and then only, we find that our state has changed. The truth is that we change without ceasing, and that the state itself is nothing but change.

This amounts to saying that there is no essential difference between passing from one state to another and persisting in the same state. If the state which "remains the same" is more varied than we think, on the other hand the passing from one state to another resembles, more than we imagine, a single state being prolonged; the transition is continuous. But, just because we close our eyes to the unceasing variation of every psychical state, we are obliged, when the change has become so considerable as to force itself on our attention, to speak as if a new state were placed alongside the previous one. Of this new state we assume that it remains unvarying in its turn, and so on endlessly. The apparent discontinuity of the psychical life is then due to our attention being fixed on it by a series of separate acts: actually there is only a gentle slope; but in following the broken line of our acts of attention, we think we perceive separate steps. True, our psychic life is full of the unforeseen. A thousand incidents arise, which seem to be cut off from those which precede them, and to be disconnected from those which follow. Discontinuous though they appear, however, in point of fact they stand out against the continuity of a background on which they are designed, and to which indeed they owe the intervals that separate them; they are the beats of the drum which break forth here and there in the symphony. Our attention fixes on them because they interest it more, but each of them is borne by the fluid mass of our whole psychical existence. Each is only the best illuminated point of a moving zone which comprises all that we feel or think or will—all, in short, that we are at any given moment. It is this entire zone which in reality makes up our state. Now, states thus defined cannot be regarded as distinct elements. They continue each other in an endless flow.

But, as our attention has distinguished and separated them artificially, it is obliged next to reunite them by an artificial bond. It imagines, therefore, a formless *ego*, indifferent and unchangeable, on which it threads the psychic states which it has set up as independent entities. Instead of a flux of fleeting shades merging into each other, it perceives distinct and, so to speak, *solid* colors, set side by side like the beads of a necklace; it must perforce then suppose a thread, also itself solid, to hold the beads together. But if this colorless substratum is perpetually colored by that which covers it, it is for us, in its indeterminateness, as if it did not exist, since we only perceive what is colored, or, in other words, psychic states. As a matter of fact, this substratum has no reality; it is merely a symbol intended to recall unceasingly to our consciousness the artificial character of the process by which the attention places clean-cut states side by side, where actually there is a continuity which unfolds. If our existence were composed of separate states with an impassive ego to unite them, for us there would be no duration. For an ego which does not change does not *endure*, and a psychic state which remains the same so long as it is not replaced by the following state does not *endure* either. Vain, therefore, is the attempt to range such states beside each other on the ego supposed to sustain them: never can these solids strung upon a solid make up that duration which flows. What we actually obtain in this way is an artificial imitation of the internal life, a static equivalent which well lend itself better to the requirements of logic and language, just because we have eliminated from it the element of real time. But, as regards the psychical life unfolding beneath the symbols which conceal it, we readily perceive that time is just the stuff it is made of.

There is, moreover, no stuff more resistant nor more substantial. For our duration is not merely one instant replacing another; if it were, there would never be anything but the present—no prolonging of the past into the actual, no evolution, no concrete duration. Duration is the continuous progress of the past which gnaws into the future and which swells as it advances. And as the past grows without ceasing, so also there is no limit to its preservation. Memory, as we have tried to prove,[1] is not a faculty of putting away recollections in a drawer, or of inscribing them in a register. There is no register, no drawer; there is not even, properly speaking, a faculty, for a faculty works intermittently, when it will or when it can, whilst the piling up of the past upon the past goes on without relaxation. In reality, the past is preserved by itself, automatically. In its entirety, probably, it follows us at every instant; all that we have felt, thought and willed from our earliest infancy is there, leaning over the present which is about to join it, pressing against the portals of consciousness that would fain leave it outside. The cerebral mechanism is arranged just so as to drive back into the unconscious almost the whole of this past, and to admit beyond the threshold only that which can cast light on the present situation or further the action now being prepared—in short, only that which can give *useful* work. At the most, a few superfluous recollections may succeed in smuggling themselves through the half-open door. These memories, messengers from the unconscious, remind us of what we are dragging behind us unawares. But, even though we may have no distinct idea of it, we feel vaguely that our past remains present to us. What are we, in fact, what is our *character*, if

[1] *Matière et mémoire,* Paris, 1896, chaps. ii. and iii.

not the condensation of the history that we have lived from our birth—nay, even before our birth, since we bring with us prenatal dispositions? Doubtless we think with only a small part of our past, but it is with our entire past, including the original bent of our soul, that we desire, will and act. Our past, then, as a whole, is made manifest to us in its impulse; it is felt in the form of tendency, although a small part of it only is known in the form of idea.

From this survival of the past it follows that consciousness cannot go through the same state twice. The circumstances may still be the same, but they will act no longer on the same person, since they find him at a new moment of his history. Our personality, which is being built up each instant with its accumulated experience, changes without ceasing. By changing, it prevents any state, although superficially identical with another, from ever repeating it in its very depth. That is why our duration is irreversible. We could not live over again a single moment, for we should have to begin by effacing the memory of all that had followed. Even could we erase this memory from our intellect, we could not from our will.

Thus our personality shoots, grows and ripens without ceasing. Each of its moments is something new added to what was before. We may go further: it is not only something new, but something unforeseeable. Doubtless, my present state is explained by what was in me and by what was acting on me a moment ago. In analyzing it I should find no other elements. But even a superhuman intelligence would not have been able to foresee the simple indivisible form which gives to these purely abstract elements their concrete organization. For to foresee consists of projecting into the future what has been perceived in the past, or of imagining for a later time a new grouping, in a new order, of elements already perceived. But that which has never been perceived, and which is at the same time simple, is necessarily unforeseeable. Now such is the case with each of our states, regarded as a moment in a history that is gradually unfolding: it is simple, and it cannot have been already perceived, since it concentrates in its indivisibility all that has been perceived and what the present is adding to it besides. It is an original moment of a no less original history.

The finished portrait is explained by the features of the model, by the nature of the artist, by the colors spread out on the palette; but, even with the knowledge of what explains it, no one, not even the artist, could have foreseen exactly what the portrait would be, for to predict it would have been to produce it before it was produced—an absurd hypothesis which is its own refutation. Even so with regard to the moments of our life, of which we are the artisans. Each of them is a kind of creation. And just as the talent of the painter is formed or deformed—in any case, is modified—under the very influence of the works he produces, so each of our states, at the moment of its issue, modifies our personality, being indeed the new form that we are just assuming. It is then right to say that what we do depends on what we are; but it is necessary to add also that we are, to a certain extent, what we do, and that we are creating ourselves continually. This creation of self by self is the more complete, the more one reasons on what one does. For reason does not proceed in such matters as in geometry, where impersonal premises are given once for all, and an impersonal conclusion must perforce be drawn. Here, on the contrary, the same reasons may dictate to different persons, or to the same person at different moments, acts profoundly different, although equally reasonable. The truth is that they are not quite the same reasons, since they are not those of the same person, nor of the same moment. That is why we cannot deal with them in the abstract, from outside, as in geometry, nor solve for another the problems by which he is faced in life. Each must solve them from within, on his own account. But we need not go more deeply into this. We are seeking only the precise meaning that our consciousness gives to this word "exist," and we find that, for a conscious being, to exist is to change, to change is to mature, to mature is to go on creating oneself endlessly. Should the same be said of existence in general?

A material object, of whatever kind, presents opposite characters to those which we have just been describing. Either it remains as it is, or else, if it changes under the influence of an external force, our idea of this change is that of a displacement of parts which themselves do not change. If these parts took to changing, we should split them up in their turn. We should thus descend to the molecules of which the fragments are made, to the atoms that make up the molecules, to the corpuscles that generate the atoms, to the "imponderable" within which the corpuscle is perhaps a mere vortex. In short, we should push the division or analysis as far as necessary. But we should stop only before the unchangeable.

Now, we say that a composite object changes by the displacement of its parts. But when a part has left its position, there is nothing to prevent its return to it. A group of elements which has gone through a state can therefore always find its way back to that state, if not by itself, at least by means of an external cause able to restore everything to its place. This amounts to saying that any state of the group may be repeated as often as desired, and consequently that the group does not grow old. It has no history.

Thus nothing is created therein, neither form nor matter. What the group will be is already present in what it is, provided "what it is" includes all the points of the universe with which it is related. A superhuman intellect could calculate, for any moment of time, the position of any point of the system in space. And as there is nothing more in the form of the whole than the arrangement of its parts, the future forms of the system are theoretically visible in its present configuration.

All our belief in objects, all our operations on the systems that science isolates, rest in fact on the idea that time does not bite into them. We have touched on this question in an earlier work, and shall return to it in the course of the present study. For the moment, we will confine ourselves to pointing out that the abstract time t attributed by science to a material object or to an isolated system consists only in a certain number of simultaneities or more generally of correspondences, and that this number remains the same, whatever be the nature of the intervals between the correspondences. With these intervals we are never concerned when dealing with inert matter; or, if they are considered, it is in order to count therein fresh correspondences, between which again we shall not care what happens. Common sense, which is occupied with detached objects, and also science, which considers isolated systems, are concerned only with the ends of the intervals and not with the intervals themselves. Therefore the flow of time might assume an infinite rapidity, the entire past, present, and future of material objects or of isolated systems might be spread out all at once in space, without there being anything to change either in the formulae of the scientist or even in the language of common sense. The number *t* would always stand for the same thing; it would still count the same number of correspondences between the states of the objects or systems and the points of the line, ready drawn, which would be then the "course of time."

Yet succession is an undeniable fact, even in the material world. Though our reasoning on isolated systems may imply that their history, past, present and future, might be instantaneously unfurled like a fan, this history, in point of fact, unfolds itself gradually, as if it occupied a duration like our own. If I want to mix a glass of sugar and water, I must, willy-nilly, wait until the sugar melts. This little fact is big with meaning. For here the time I have to wait is not that mathematical time which would apply equally well to the entire history of the material world, even if that history were spread out instantaneously in space. It coincides with my impatience, that is to say, with a certain portion of my own duration, which I cannot protract or contract as I like. It is no longer something *thought*, it is something *lived*. It is no longer a relation, it is an absolute. What else can this mean than that the glass of water, the sugar, and the process of the sugar's melting in the water are abstractions, and that the Whole within which they have been cut out by my senses and understanding progresses, it may be in the manner of a consciousness?

Certainly, the operation by which science isolates and closes a system is not altogether artificial. If it had no objective foundation, we could not explain why it is clearly indicated in some cases and impossible in others. We shall see that matter has a tendency to constitute *isolable* systems, that can be treated geometrically. In fact, we shall define matter by just this tendency. But it is only a tendency. Matter does not go to the end, and the isolation is never complete. If science does go to the end and isolate completely, it is for convenience of study; it is understood that the so-called isolated system remains subject to certain external influences. Science merely leaves these alone, either because it finds them slight enough to be negligible, or because it intends to take them into account later on. It is none the less true that these influences are so many threads which bind up the system to another more extensive, and to this a third which includes both, and so on to the system most objectively isolated and most independent of all, the solar system complete. But, even here, the isolation is not absolute. Our sun radiates heat and light beyond the farthest planet. And, on the other hand, it moves in a certain fixed direction, drawing with it the planets and their satellites. The thread attaching it to the rest of the universe is doubtless very tenuous. Nevertheless it is along this thread that is transmitted down to the smallest particle of the world in which we live the duration immanent to the whole of the universe.

The universe *endures*. The more we study the nature of time, the more we shall comprehend that duration means invention, the creation of forms, the continual elaboration of the absolutely new. The systems marked off by science *endure* only because they are bound up inseparably with the rest of the universe. It is true that in the universe itself two opposite movements are to be distinguished, as we shall see later on, "descent" and "ascent." The first only unwinds a roll ready prepared. In principle, it might be accomplished almost instantaneously, like releasing a spring. But the ascending movement, which corresponds to an inner work of ripening or creating, *endures* essentially, and imposes its rhythm on the first, which is inseparable from it.

There is no reason, therefore, why a duration, and so a form of existence like our own, should not be attributed to the systems that science isolates, provided such systems are reintegrated into the Whole. But they must be so reintegrated. The same is even more obviously true of the objects cut out by our perception. The distinct outlines which we see in an object, and which give it its individuality, are only the design of a certain kind of *influence* that we might exert on a certain point of space: it is the plan of our eventual actions that is sent back to our eyes, as though by a mirror, when we see the surfaces and edges of things. Suppress this action, and with it consequently those main directions which by perception are traced out for it in the entanglement of the real, and the individuality of the body is reabsorbed in the universal interaction which, without doubt, is reality itself.

Now, we have considered material objects generally. Are there not some objects privileged? The bodies we perceive are, so to speak, cut out of the stuff of nature by our *perception*, and the scissors follow, in some way, the marking of lines along which *action* might be taken. But the body which is to perform this action, the body which marks out upon matter the design of its eventual actions even before they are actual, the body that has only to point its sensory organs on the flow of the real in order to make that flow crystallize into definite forms and thus to create all the other bodies—in short, the *living* body—is this a body as others are?

Doubtless it, also, consists in a portion of extension bound up with the rest of extension, an intimate part of the Whole, subject to the same physical and chemical laws that govern any and every portion of matter. But, while the subdivision of matter into separate bodies is relative to our perception, while the building up of closed-off systems of material points is relative to our science, the living body has been separated and closed off by nature herself. It is composed of unlike parts that complete each other. It performs diverse functions that involve each other. It is an *individual*, and of no other object, not even of the crystal, can this be said, for a crystal has neither difference of parts nor diversity of functions. No doubt, it is hard to decide, even in the organized world, what is individual and what is not.

The difficulty is great, even in the animal kingdom; with plants it is almost insurmountable. This difficulty is, moreover, due to profound causes, on which we shall dwell later. We shall see that individuality admits of any number of degrees, and that it is not fully realized anywhere, even in man. But that is no reason for thinking it is not a characteristic property of life. The biologist who proceeds as a geometrician is too ready to take advantage here of our inability to give a precise and general definition of individuality. A perfect definition applies only to a *completed* reality; now, vital properties are never entirely realized, though always on the way to become so; they are not so much *states* as *tendencies*. And a tendency achieves all that it aims at only if it is not thwarted by another tendency. How, then, could this occur in the domain of life, where, as we shall show, the interaction of antagonistic tendencies is always implied? In particular, it may be said of individuality that, while the tendency to individuate is everywhere present in the organized world, it is everywhere opposed by the tendency toward reproduction. For the individuality to be perfect, it would be necessary that no detached part of the organism could live separately. But then reproduction would be impossible. For what is reproduction, but the building up of a new organism with a detached fragment of the old? Individuality therefore harbors its enemy at home. Its very need of perpetuating itself in time condemns it never to be complete in space. The biologist must take due account of both tendencies in every instance, and it is therefore useless to ask him for a definition of individuality that shall fit all cases and work automatically.

But too often one reasons about the things of life in the same way as about the conditions of crude matter. Nowhere is the confusion so evident as in discussions about individuality. We are shown the stumps of a Lumbriculus, each regenerating its head and living thenceforward as an independent individual; a hydra whose pieces become so many fresh hydras; a sea-urchin's egg whose fragments develop complete embryos: where then, we are asked, was the individuality of the egg, the hydra, the worm?—But, because there are several individuals now, it does not follow that there was not a single individual just before. No doubt, when I have seen several drawers fall from a chest, I have no longer the right to say that the article was all of one piece. But the fact is that there can be nothing more in the present of the chest of drawers than there was in its past, and if it is made up of several different pieces now, it was so from the date of its manufacture. Generally speaking, unorganized bodies, which are what we have need of in order that we may act, and on which we have modeled our fashion of thinking, are regulated by this simple law: *the present contains nothing more than the past, and what is found in the effect was already in the cause.* But suppose that the distinctive feature of the organized body is that it grows and changes without ceasing, as indeed the most superficial observation testifies, there would be nothing astonishing in the fact that it was *one* in the first instance, and afterwards *many*. The reproduction of unicellular organisms consists in just this—the living being divides into two halves, of which each is a complete individual. True, in the more complex animals, nature localizes in the almost independent sexual cells the power of producing the whole anew. But something of its power may remain diffused in the rest of the organism, as the facts of regeneration prove, and it is conceivable that in certain privileged cases the faculty may persist integrally in a latent condition and manifest itself on the first opportunity. In truth, that I may have the right to speak of individuality, it is not necessary that the organism should be without the power to divide into fragments that are able to live. It is sufficient that it should have presented a certain systematization of parts before the division, and that the same systematization tend to be reproduced in each separate portion afterwards. Now, that is precisely what we observe in the organic world. We may conclude, then, that individuality is never perfect, and that it is often difficult, sometimes impossible, to tell what is an individual, and what is not, but that life nevertheless manifests a search for individuality, as if it strove to constitute systems naturally isolated, naturally closed.

By this is a living being distinguished from all that our perception or our science isolates or closes artificially. It would therefore be wrong to compare it to an *object*. Should we wish to find a term of comparison in the inorganic world, it is not to a determinate material object, but much rather to the totality of the material universe that we ought to compare the living organism. It is true that the comparison would not be worth much, for a living being is observable, whilst the whole of the universe is constructed or reconstructed by thought. But at least our attention would thus have been called to the essential character of organization. Like the universe as a whole, like each conscious being taken separately, the organism which lives is a thing that *endures*. Its past, in its entirety, is prolonged into its present, and abides there, actual and acting. How otherwise could we understand that it passes through distinct and well-marked phases, that it changes its age—in short, that it has a history? If I consider my body in particular, I find that, like my consciousness, it matures little by little from infancy to old age; like myself, it grows old. Indeed, maturity and old age are, properly speaking, attributes only of my body; it is only metaphorically that I apply the same names to the corresponding changes of my conscious self. Now, if I pass from the top to the bottom of the scale of living beings, from one of the most to one of the least differentiated, from the multicellular organism of man to the unicellular organism of the Infusorian, I find, even in this simple cell, the same process of growing old. The Infusorian is exhausted at the end of a certain number of divisions, and though it may be possible, by modifying the environment, to put off the moment when a rejuvenation by conjugation becomes necessary, this cannot be indefinitely postponed.[1] It is true that between these two extreme cases, in which the organism is completely individualized, there might be found a multitude of others in which the individuality is less well marked, and in which, although there is doubtless an aging somewhere, one cannot say exactly what it is that grows old. Once more, there is no universal biological law which applies

[1] Calkins, *Studies on the Life History of Protozoa* (*Archiv f. Entwicklungsmechanik*, vol. xv., 1903, pp. 139-186).

precisely and automatically to every living thing. There are only *directions* in which life throws out species in general. Each particular species, in the very act by which it is constituted, affirms its independence, follows its caprice, deviates more or less from the straight line, sometimes even remounts the slope and seems to turn its back on its original direction. It is easy enough to argue that a tree never grows old, since the tips of its branches are always equally young, always equally capable of engendering new trees by budding. But in such an organism—which is, after all, a society rather than an individual—*something* ages, if only the leaves and the interior of the trunk. And each cell, considered separately, evolves in a specific way. *Wherever anything lives, there is, open somewhere, a register in which time is being inscribed.*

This, it will be said, is only a metaphor.—It is of the very essence of mechanism, in fact, to consider as metaphorical every expression which attributes to time an effective action and a reality of its own. In vain does immediate experience show us that the very basis of our conscious existence is memory, that is to say, the prolongation of the past into the present, or, in a word, *duration,* acting and irreversible. In vain does reason prove to us that the more we get away from the objects cut out and the systems isolated by common sense and by science and the deeper we dig beneath them, the more we have to do with a reality which changes as a whole in its inmost states, as if an accumulative memory of the past made it impossible to go back again. The mechanistic instinct of the mind is stronger than reason, stronger than immediate experience. The metaphysician that we each carry unconsciously within us, and the presence of which is explained, as we shall see later on, by the very place that man occupies amongst the living beings, has its fixed requirements, its ready-made explanations, its irreducible propositions: all unite in denying concrete duration. Change *must* be reducible to an arrangement or rearrangement of parts; the irreversibility of time *must* be an appearance relative to our ignorance; the impossibility of turning back *must* be only the inability of man to put things in place again. So growing old can be nothing more than the gradual gain or loss of certain substances, perhaps both together. Time is assumed to have just as much reality for a living being as for an hour-glass, in which the top part empties while the lower fills, and all goes where it was before when you turn the glass upside down.

True, biologists are not agreed on what is gained and what is lost between the day of birth and the day of death. There are those who hold to the continual growth in the volume of protoplasm from the birth of the cell right on to its death.[1] More probable and more profound is the theory according to which the diminution bears on the quantity of nutritive substance contained in that "inner environment" in which the organism is being renewed, and the increase on the quantity of unexcreted residual substances which, accumulating in the body, finally "crust it over." [2] Must we however—with

[1] Sedgwick Minot, *On Certain Phenomena of Growing Old* (*Proc. Amer. Assoc. for the Advancement of Science*, 39th Meeting, Salem, 1891, pp. 271-288).

[2] Le Dantec, *L'Individualité et l'erreur individualiste,* Paris, 1905, pp 84 ff.

an eminent bacteriologist—declare any explanation of growing old insufficient that does not take account of phagocytosis? [1] We do not feel qualified to settle the question. But the fact that the two theories agree in affirming the constant accumulation or loss of a certain kind of matter, even though they have little in common as to what is gained and lost, shows pretty well that the frame of the explanation has been furnished *a priori.* We shall see this more and more as we proceed with our study: it is not easy, in thinking of time, to escape the image of the hour-glass.

The cause of growing old must lie deeper. We hold that there is unbroken continuity between the evolution of the embryo and that of the complete organism. The impetus which causes a living being to grow larger, to develop and to age, is the same that has caused it to pass through the phases of the embryonic life. The development of the embryo is a perpetual change of form. Anyone who attempts to note all its successive aspects becomes lost in an infinity, as is inevitable in dealing with a continuum. Life does but prolong this prenatal evolution. The proof of this is that it is often impossible for us to say whether we are dealing with an organism growing old or with an embryo continuing to evolve; such is the case, for example, with the larvae of insects and crustacea. On the other hand, in an organism such as our own, crises like puberty or the menopause, in which the individual is completely transformed, are quite comparable to changes in the course of larval or

[1] Metchnikoff, *La Dégénérescence sénile* (*Année biologique*, iii., 1897, pp. 249 ff.). Cf. by the same author, *La Nature humaine*, Paris, 1903, pp. 312 ff.

embryonic life—yet they are part and parcel of the process of our aging. Although they occur at a definite age and within a time that may be quite short, no one would maintain that they appear then *ex abrupto,* from without, simply because a certain age is reached, just as a legal right is granted to us on our one-and-twentieth birthday. It is evident that a change like that of puberty is in course of preparation at every instant from birth, and even before birth, and that the aging up to that crisis consists, in part at least, of this gradual preparation. In short, what is properly vital in growing old is the insensible, infinitely graduated, continuance of the change of form. Now, this change is undoubtedly accompanied by phenomena of organic destruction: to these, and to these alone, will a mechanistic explanation of aging be confined. It will note the facts of sclerosis, the gradual accumulation of residual substances, the growing hypertrophy of the protoplasm of the cell. But under these visible effects an inner cause lies hidden. The evolution of the living being, like that of the embryo, implies a continual recording of duration, a persistence of the past in the present, and so an appearance, at least, of organic memory.

The present state of an unorganized body depends exclusively on what happened at the previous instant; and likewise the position of the material points of a system defined and isolated by science is determined by the position of these same points at the moment immediately before. In other words, the laws that govern unorganized matter are expressible, in principle, by differential equations in which time (in the sense in which the mathematician takes this word) would play the role of independent variable. Is it so with the laws of life? Does the state of a living body find its complete explanation in the state immediately before? Yes, if it is agreed *a priori* to liken the living body to other bodies, and to identify it, for the sake of the argument, with the artificial systems on which the chemist, physicist and astronomer operate. But in astronomy, physics and chemistry the proposition has a perfectly definite meaning: it signifies that certain aspects of the present, important for science, are calculable as functions of the immediate past. Nothing of the sort in the domain of life. Here calculation touches, at most, certain phenomena of organic *destruction*. Organic *creation*, on the contrary, the evolutionary phenomena which properly constitute life, we cannot in any way subject to a mathematical treatment. It will be said that this impotence is due only to our ignorance. But it may equally well express the fact that the present moment of a living body does not find its explanation in the moment immediately before, that *all* the past of the organism must be added to that moment, its heredity—in fact, the whole of a very long history. In the second of these two hypotheses, not in the first, is really expressed the present state of the biological sciences, as well as their direction. As for the idea that the living body might be treated by some superhuman calculator in the same mathematical way as our solar system, this has gradually arisen from a metaphysic which has taken a more precise form since the physical discoveries of Galileo, but which, as we shall show, was always the natural metaphysic of the human mind. Its apparent clearness, our impatient desire to find it true, the enthusiasm with which so many excellent minds accept it without proof —all the seductions, in short, that it exercises on our thought, should put us on our guard against it. The attraction it has for us proves well enough that it gives satisfaction to an innate inclination. But, as will be seen further on, the intellectual tendencies innate today, which life must have created in the course of its evolution, are not at all meant to supply us with an explanation of life: they have something else to do.

Any attempt to distinguish between an artificial and a natural system, between the dead and the living, runs counter to this tendency at once. Thus it happens that we find it equally difficult to imagine that the organized has duration and that the unorganized has not. When we say that the state of an artificial system depends exclusively on its state at the moment before, does it not seem as if we were bringing time in, as if the system had something to do with real duration? And, on the other hand, though the whole of the past goes into the making of the living being's present moment, does not organic memory press it into the moment immediately before the present, so that the moment immediately before becomes the sole cause of the present one?—To speak thus is to ignore the cardinal difference between *concrete* time, along which a real system develops, and that *abstract* time which enters into our speculations on artificial systems. What does it mean, to say that the state of an artificial system depends on what it was at the moment immediately before? There is no instant immediately before another instant; there could not be, any more than there could be one mathematical point touching another. The instant "immediately before" is, in reality, that which is connected with the present instant by the interval *dt*. All that you mean to say, therefore, is that the present state of the system is defined by equations into which differential coefficients enter, such as $ds|dt$, $dv|dt$, that is to say, at bottom, *present* velocities and *present* accelerations. You are therefore really speaking only of the present—a present, it is true, considered along with its *tendency*. The systems science works with are, in fact, in an instantaneous present that is always being renewed; such systems are never in that real, concrete duration in which the past remains bound up with the present. When the mathematician calculates the future state of a system at the end of a time *t*, there is nothing to prevent him from supposing that the universe vanishes from this moment till that, and suddenly reappears. It is the *t*-th moment only that counts—and that will be a mere instant. What will flow on in the interval—that is to say, real time—does not count, and cannot enter into the calculation. If the mathematician says that he puts himself inside this interval, he means that he is placing himself at a certain point, at a particular moment, therefore at the extremity again of a certain time t'; with the interval up to T' he is not concerned. If he divides the interval into infinitely small parts by considering the differential *dt*, he thereby expresses merely the fact that he will consider accelerations and velocities—that is to say, numbers which denote tendencies and enable him to calculate the state of the system at a given moment. But he is always speaking of a given moment—a static moment, that is—and not of flowing time. In short, *the world the mathematician deals with is a world that dies and is reborn at every instant—the world which Descartes was thinking of when he spoke of continued creation.* But, in time thus conceived, how could evolution, which is the very essence of life, ever take place? Evolution implies a real persistence of the past in the present, a duration which is, as it were, a hyphen, a connecting link. In other words, to know a living being or *natural system* is to get at the very interval of duration, while the knowledge of an *artificial* or *mathematical system* applies only to the extremity.

Continuity of change, preservation of the past in the present, real duration—the living being seems, then, to share these attributes with consciousness. Can we go further and say that life, like conscious activity, is invention, is unceasing creation?

It does not enter into our plan to set down here the proofs of transformism. We wish only to explain in a word or two why we shall accept it, in the present work, as a sufficiently exact and precise expression of the facts actually known. The idea of transformism is already in germ in the natural classification of organized beings. The naturalist, in fact, brings together the organisms that are like each other, then divides the group into sub-groups within which the likeness is still greater, and so on: all through the operation, the characters of the group appear as general themes on which each of the sub-groups performs its particular variation. Now, such is just the relation we find, in the animal and in the vegetable world between the generator and the generated: on the canvas which the ancestor passes on, and which his descendants possess in common, each puts his own original embroidery. True, the differences between the descendant and the ancestor are slight, and it may be asked whether the same living matter presents enough plasticity to take in turn such different forms as those of a fish, a reptile and a bird. But, to this question, observation gives a peremptory answer. It shows that up to a certain period in its development the embryo of the bird is hardly distinguishable from that of the reptile, and that the individual develops, throughout the embryonic life in general, a series of transformations comparable to those through which, according to the theory of evolution, one species passes into another. A single cell, the result of the combination of two cells, male and female, accomplishes this work by dividing. Every day, before our eyes, the highest forms of life are springing from a very elementary form. Experience, then, shows that the most complex has been able to issue from the most simple by way of evolution. Now, has it arisen so, as a matter of fact? Paleontology, in spite of the insufficiency of its evidence, invites us to believe it has; for, where it makes out the order of succession of species with any precision, this order is just what considerations drawn from embryogeny and comparative anatomy would lead anyone to suppose, and each new paleontological discovery brings transformism a new confirmation. Thus, the proof drawn from mere observation is ever being strengthened, while, on the other hand, experiment is removing the objections one by one. The recent experiments of H. de Vries, for instance, by showing that important variations can be produced suddenly and transmitted regularly, have overthrown some of the greatest difficulties raised by the theory. They have enabled us greatly to shorten the time biological evolution seems to demand. They also render us less exacting toward paleontology. So that, all things considered, the transformist hypothesis looks more and more like a close approximation to the truth. It is not rigorously demonstrable; but, failing the certainty of theoretical or experimental demonstration, there is a probability which is continually growing, due to evidence which, while coming short of direct proof, seems to point persistently in its direction: such is the kind of probability that the theory of transformism offers.

Let us admit, however, that transformism may be wrong. Let us suppose that species are proved, by inference or by experiment, to have arisen by a discontinuous process, of which today we have no idea. Would the doctrine be affected in so far as it has a special interest or importance for us? Classification would probably remain, in its broad lines. The actual data of embryology would also remain. The correspondence between comparative embryogeny and comparative anatomy would remain too. Therefore biology could and would continue to establish between living forms the same relations and the same kinship as transformism supposes today. It would be, it is true, an *ideal* kinship, and no longer a *material* affiliation. But, as the actual data of paleontology would also remain, we should still have to admit that it is successively, not simultaneously, that the forms between which we find an ideal kinship have appeared. Now, the evolutionist theory, so far as it has any importance for philosophy, requires no more. It consists above all in establishing relations of ideal kinship, and in maintaining that wherever there is this relation of, so to speak, *logical* affiliation between forms, there is also a relation of *chronological* succession between the species in which these forms are materialized. Both arguments would hold in any case. And hence, an evolution *somewhere* would still have to be supposed, whether in a creative Thought in which the ideas of the different species are generated by each other exactly as transformism holds that species themselves are generated on the earth; or in a plan of vital organization immanent in nature, which gradually works itself out, in which the relations of logical and chronological affiliation between pure forms are just those which transformism presents as relations of real affiliation between living individuals; or, finally, in some unknown cause of life, which develops its effects *as if* they generated one another. Evolution would then simply have been *transposed,* made to pass from the visible to the invisible. Almost all that transformism tells us today would be preserved, open to interpretation in another way. Will it not, therefore, be better to stick to the letter of transformism as almost all scientists profess it? Apart from the question to what extent the theory of evolution describes the facts and to what extent it symbolizes them, there is nothing in it that is irreconcilable with the doctrines it has claimed to replace, even with that of special creations, to which it is usually opposed. For this reason we think the language of transformism forces itself now upon all philosophy, as the dogmatic affirmation of transformism forces itself upon science.

But then, we must no longer speak of *life in general* as an abstraction, or as a mere heading under which all living beings are inscribed. At a certain moment, in certain points of space, a visible current has taken rise; this current of life, traversing the bodies it has organized one after another, passing from generation to generation, has become divided amongst species and distributed amongst individuals without losing anything of its force, rather intensifying in proportion to its advance. It is well known that, on the theory of the "continuity of the germ-plasm," maintained by Weismann, the sexual elements of the generating organism pass on their properties directly to the sexual elements of the organism engendered. In this extreme form, the theory has seemed debatable, for it is only in exceptional cases that there are any signs of sexual glands at the time of segmentation of the fertilized egg. But, though the cells that engender the sexual elements do not generally appear at the beginning of the embryonic life, it is none the less true that they are always formed out of those tissues of the embryo which have not undergone any particular functional differentiation, and whose cells are made of unmodified protoplasm.[1] In other words, the genetic power of the fertilized ovum weakens, the more it is spread over the growing mass of the tissues of the embryo; but, while it is being thus diluted, it is concentrating anew something of itself on a certain special point, to wit, the cells, from which the ova or spermatozoa will develop. It might therefore be said that, though the germ-plasm is not continuous, there is at least continuity of genetic energy, this energy being expended only at certain instants, for just enough time to give the requisite impulsion to the embryonic life, and

[1] Roule, *L'Embryologie générale*, Paris, 1893, p. 319.

being recouped as soon as possible in new sexual elements, in which, again, it bides its time. Regarded from this point of view, *life is like a current passing from germ to germ through the medium of a developed organism*. It is as if the organism itself were only an excrescence, a bud caused to sprout by the former germ endeavoring to continue itself in a new germ. The essential thing is the *continuous progress* indefinitely pursued, an invisible progress, on which each visible organism rides during the short interval of time given it to live.

Now, the more we fix our attention on this continuity of life, the more we see that organic evolution resembles the evolution of a consciousness, in which the past presses against the present and causes the upspringing of a new form of consciousness, incommensurable with its antecedents. That the appearance of a vegetable or animal species is due to specific causes, nobody will gainsay. But this can only mean that if, after the fact, we could know these causes in detail, we could explain by them the form that has been produced; foreseeing the form is out of the question.[1] It may perhaps be said that the form could be foreseen if we could know, in all their details, the conditions under which it will be produced. But these conditions are built up into it and are part and parcel of its being; they are peculiar to that phase of its history in which life finds itself at the moment of producing the form: how could we know be-

[1] The irreversibility of the series of living beings has been well set forth by Baldwin (*Development and Evolution*, New York, 1902; in particular p. 327).

forehand a situation that is unique of its kind, that has never yet occurred and will never occur again? Of the future, only that is foreseen which is like the past or can be made up again with elements like those of the past. Such is the case with astronomical, physical and chemical facts, with all facts which form part of a system in which elements supposed to be unchanging are merely put together, in which the only changes are changes of position, in which there is no theoretical absurdity in imagining that things are restored to their place; in which, consequently, the same total phenomenon, or at least the same elementary phenomena, can be repeated. But an original situation, which imparts something of its own originality to its elements, that is to say, to the partial views that are taken of it, how can such a situation be pictured as given before it is actually produced? [1] All that can be said is that, once produced, it will be explained by the elements that analysis will then carve out of it. Now, what is true of the production of a new species is also true of the production of a new individual, and, more generally, of any moment of any living form. For, though the variation must reach a certain importance and a certain generality in order to give rise to a new species, it is being produced every moment, continuously and insensibly, in every living being. And it is evident that even the sudden "mutations" which we now hear of are possible only if a process of incubation, or rather of maturing, is going on throughout a series of generations that do not seem

[1] We have dwelt on this point and tried to make it clear in the *Essai sur les données immédiates de la conscience*, pp. 140-151.

to change. In this sense it might be said of life, as of consciousness, that at every moment it is creating something.[1]

But against this idea of the absolute originality and unforeseeability of forms our whole intellect rises in revolt. The essential function of our intellect, as the evolution of life has fashioned it, is to be a light for our conduct, to make ready for our action on things, to foresee, for a given situation, the events, favorable or unfavorable, which may follow thereupon. Intellect therefore instinctively selects in a given situation whatever is like something already known; it seeks this out, in order that it may apply its principle that "like produces like." In just this does the prevision of the future by common sense consist. Science carries this faculty to the highest possible degree of exactitude and precision, but does not alter its essential character. Like ordinary knowledge, in dealing with things science is concerned only with the aspect of *repetition*. Though the whole be original, science will always manage to analyze it into elements or aspects which are approximately a reproduction of the past. Science can work only on what is supposed to re-

[1] In his fine work on *Genius in Art* (*Le Génie dans l'art*), M. Séailles develops this twofold thesis, that art is a continuation of nature and that life is creation. We should willingly accept the second formula; but by creation must we understand, as the author does, a *synthesis* of elements? Where the elements pre-exist, the synthesis that will be made is virtually given, being only one of the possible arrangements. This arrangement a superhuman intellect could have perceived in advance among all the possible ones that surround it. We hold, on the contrary, that in the domain of life the elements have no real and separate existence. They are manifold mental views of an indivisible process. And for that reason there is radical contingency in progress, incommensurability between what goes before and what follows—in short duration.

peat itself—that is to say, on what is withdrawn, by hypothesis, from the action of real time. Anything that is irreducible and irreversible in the successive moments of a history eludes science. To get a notion of this irreducibility and irreversibility, we must break with scientific habits which are adapted to the fundamental requirements of thought, we must do violence to the mind, go counter to the natural bent of the intellect. But that is just the function of philosophy.

In vain, therefore, does life evolve before our eyes as a continuous creation of unforeseeable form: the idea always persists that form, unforeseeability and continuity are mere appearance—the outward reflection of our own ignorance. What is presented to the senses as a continuous history would break up, we are told, into a series of successive states. "What gives you the impression of an original state resolves, upon analysis, into elementary facts, each of which is the repetition of a fact already known. What you call an unforeseeable form is only a new arrangement of old elements. The elementary causes, which in their totality have determined this arrangement, are themselves old causes repeated in a new order. Knowledge of the elements and of the elementary causes would have made it possible to foretell the living form which is their sum and their resultant. When we have resolved the biological aspect of phenomena into physico-chemical factors, we will leap, if necessary, over physics and chemistry themselves; we will go from masses to molecules, from molecules to atoms, from atoms to corpuscles: we must indeed at last come to something that can be treated as a kind of solar system, astronomically. If you deny it, you oppose the very principle of scientific mechanism, and you arbitrarily affirm that living matter is not made of the same elements as other matter."—We reply that we do not question the fundamental identity of inert matter and organized matter. The only question is whether the natural systems which we call living beings must be assimilated to the artificial systems that science cuts out within inert matter, or whether they must not rather be compared to that natural system which is the whole of the universe. That life is a kind of mechanism I cordially agree. But is it the mechanism of parts artificially isolated within the whole of the universe, or is it the mechanism of the real whole? The real whole might well be, we conceive, an indivisible continuity. The systems we cut out within it would, properly speaking, not then be *parts* at all; they would be *partial views* of the whole. And, with these partial views put end to end, you will not make even a beginning of the reconstruction of the whole, any more than, by multiplying photographs of an object in a thousand different aspects, you will reproduce the object itself. So of life and of the physico-chemical phenomena to which you endeavor to reduce it. Analysis will undoubtedly resolve the process of organic creation into an ever-growing number of physico-chemical phenomena, and chemists and physicists will have to do, of course, with nothing but these. But it does not follow that chemistry and physics will ever give us the key to life.

A very small element of a curve is very near being a straight line. And the smaller it is, the nearer. In the limit, it may be termed a part of the curve or a part of the straight line, as you please, for in each of its points a curve coincides with its tangent. So likewise "vitality" is tangent, at any and every point, to physical and chemical forces; but such points are, as a fact, only views taken by a mind which imagines stops at various moments of the movement that generates the curve. In reality, life is no more made of physico-chemical elements than a curve is composed of straight lines.

In a general way, the most radical progress a science can achieve is the working of the completed results into a new scheme of the whole, by relation to which they become instantaneous and motionless views taken at intervals along the continuity of a movement. Such, for example, is the relation of modern to ancient geometry. The latter, purely static, worked with figures drawn once for all; the former studies the varying of a function—that is, the continuous movement by which the figure is described. No doubt, for greater strictness, all considerations of motion may be eliminated from mathematical processes; but the introduction of motion into the genesis of figures is nevertheless the origin of modern mathematics. We believe that if biology could ever get as close to its object as mathematics does to its own, it would become, to the physics and chemistry of organized bodies, what the mathematics of the moderns has proved to be in relation to ancient geometry. The wholly superficial displacements of masses and molecules studied in physics and chemistry would become, by relation to that inner vital movement (which is transformation and not translation) what the position of a moving object is to the movement of that object in space. And, so far as we can see, the procedure by which we should then pass from the definition of a certain vital action to the system of physico-chemical facts which it implies would be like passing from the function to its derivative, from the equation of the curve (*i.e.* the law of the continuous movement by which the curve is generated) to the equation of the tangent giving its instantaneous direction. Such a science would be a *mechanics of transformation,* of which our *mechanics of translation* would become a particular case, a simplification, a projection on the plane of pure quantity. And just as an infinity of functions have the same differential, these functions differing from each other by a constant, so perhaps the integration of the physico-chemical elements of properly vital action might determine that action only in part—a part would be left to indetermination. But such an integration can be no more than dreamed of; we do not pretend that the dream will ever be realized. We are only trying, by carrying a certain comparison as far as possible, to show up to what point our theory goes along with pure mechanism, and where they part company.

Imitation of the living by the unorganized may, however, go a good way. Not only does chemistry make organic syntheses, but we have succeeded in reproducing artificially the external appearance of certain facts of organization, such as indirect cell-division and protoplasmic circulation. It is well known that the protoplasm of the cell effects various movements within its envelope; on the other hand, indirect cell-division is the outcome of very complex operations, some involving the nucleus and others the cytoplasm. These latter commence by the doubling of the centrosome, a small spherical body alongside the nucleus. The two centrosomes thus obtained draw apart, attract the broken and doubled ends of the filament of which the original nucleus mainly consisted, and join them to form two fresh nuclei about which the two new cells are constructed which will succeed the first. Now, in their broad lines and in their external appearance, some at least of these operations have been successfully imitated. If some sugar or table salt is pulverized and some very old oil is added, and a drop of the mixture is observed under the microscope, a froth of alveolar structure is seen whose configuration is like that of protoplasm, according to certain theories, and in which movements take place which are decidedly like those of protoplasmic circulation.[1] If, in a froth of the same kind, the air is extracted from an alveolus, a cone of attraction is seen to form, like those about the centrosomes which result in the division of the nucleus.[2] Even the external motions of a unicellular organism—of an amoeba, at any rate—are sometimes explained mechanically. The displacements of an amoeba in a drop of water would be comparable to the motion to and fro of a grain of dust in a draughty room. Its mass is all the time absorbing certain soluble matters contained in the surrounding water, and giving back to it certain others; these continual exchanges, like those between two vessels separated by a porous partition, would create an ever-changing vortex around the little organism. As for the temporary prolongations or pseudopodia which the amoeba seems to make, they would be not so much given out by it as at-

[1] Bütschli, *Untersuchungen über mikroskopische Schäume und das Protoplasma*, Leipzig, 1892, First Part.

[2] Rhumbler, *Versuch einer mechanischen Erklärung der indirekten Zell- und Kernteilung* (*Roux's Archiv*, 1896).

tracted from it by a kind of inhalation or suction of the surrounding medium.[1] In the same way we may perhaps come to explain the more complex movements which the Infusorian makes with its vibratory cilia, which, moreover, are probably only fixed pseudopodia.

But scientists are far from agreed on the value of explanations and schemata of this sort. Chemists have pointed out that even in the organic—not to go so far as the organized—science has reconstructed hitherto nothing but waste products of vital activity; the peculiarly active plastic substances obstinately defy synthesis. One of the most notable naturalists of our time has insisted on the opposition of two orders of phenomena observed in living tissues, *anagenesis* and *katagenesis*. The role of the anagenetic energies is to raise the inferior energies to their own level by assimilating inorganic substances. They *construct* the tissues. On the other hand, the actual functioning of life (excepting, of course, assimilation, growth and reproduction) is of the katagenetic order, exhibiting the fall, not the rise, of energy. It is only with these facts of katagenetic order that physico-chemistry deals—that is, in short, with the dead and not with the living.[2] The other kind of facts certainly seem to defy physico-chemical analysis, even if they are not anagenetic in the proper sense of the word. As for the artificial imitation of the outward appearance of protoplasm, should a real theoretic importance be attached to this when the question of the phys-

[1] Berthold, *Studien über Protoplasmamechanik*, Leipzig, 1886, p. 102. Cf. the explanation proposed by Le Dantec, *Théorie nouvelle de la vie*, Paris, 1896, p. 60.

[2] Cope, *The Primary Factors of Organic Evolution, Chicago*, 1896, pp. 475-484.

ical framework of protoplasm is not yet settled? We are still further from compounding protoplasm chemically. Finally, a physico-chemical explanation of the motions of the amoeba, and *a fortiori* of the behavior of the Infusoria, seems impossible to many of those who have closely observed these rudimentary organisms. Even in these humblest manifestations of life they discover traces of an effective psychological activity.[1] But instructive above all is the fact that the tendency to explain everything by physics and chemistry is discouraged rather than strengthened by deep study of histological phenomena. Such is the conclusion of the truly admirable book which the histologist E. B. Wilson has devoted to the development of the cell: "The study of the cell has, on the whole, seemed to widen rather than to narrow the enormous gap that separates even the lowest forms of life from the inorganic world." [2]

To sum up, those who are concerned only with the functional activity of the living being are inclined to believe that physics and chemistry will give us the key to biological processes.[3] They have chiefly to do, as a fact, with phenomena that are *repeated* continually in the living being, as in a chemical retort. This explains,

[1] Maupas, "Etude des infusoires ciliés" (*Arch. de zoologie expérimentale*, 1883, pp. 47, 491, 518, 549, in particular). P. Vignon, *Recherches de cytologie générale sur les épithéliums*, Paris, 1902, p. 655. A profound study of the motions of the Infusoria and a very penetrating criticism of the idea of tropism have been made recently by Jennings (*Contributions to the Study of the Behavior of Lower Organisms*, Washington, 1904). The "type of behavior" of these lower organisms, as Jennings defines it (pp. 237-252), is unquestionably of the psychological order.

[2] E. B. Wilson, *The Cell in Development and Inheritance*, New York, 1897, p. 330.

[3] Dastre, *La Vie et la mort*, p. 43.

in some measure, the mechanistic tendencies of physiology. On the contrary, those whose attention is concentrated on the minute structure of living tissues, on their genesis and evolution, histologists and embryogenists on the one hand, naturalists on the other, are interested in the retort itself, not merely in its contents. They find that this retort creates its own form through a *unique* series of acts that really constitute a *history*. Thus, histologists, embryogenists, and naturalists believe far less readily than physiologists in the physico-chemical character of vital actions.

The fact is, neither one nor the other of these two theories, neither that which affirms nor that which denies the possibility of chemically producing an elementary organism, can claim the authority of experiment. They are both unverifiable, the former because science has not yet advanced a step toward the chemical synthesis of a living substance, the second because there is no conceivable way of proving experimentally the impossibility of a fact. But we have set forth the theoretical reasons which prevent us from likening the living being, a system closed off by nature, to the systems which our science isolates. These reasons have less force, we acknowledge, in the case of a rudimentary organism like the amoeba, which hardly evolves at all. But they acquire more when we consider a complex organism which goes through a regular cycle of transformations. The more duration marks the living being with its imprint, the more obviously the organism differs from a mere mechanism, over which duration glides without penetrating. And the demonstration has most force when it applies to the evolution of life as a whole, from its humblest origins to its highest forms, inasmuch as this evolution constitutes, through the unity and continuity of the animated matter which supports it, a single indivisible history. Thus viewed, the evolutionist hypothesis does not seem so closely akin to the mechanistic conception of life as it is generally supposed to be. Of this mechanistic conception we do not claim, of course, to furnish a mathematical and final refutation. But the refutation which we draw from the consideration of real time, and which is, in our opinion, the only refutation possible, becomes the more rigorous and cogent the more frankly the evolutionist hypothesis is assumed. We must dwell a good deal more on this point. But let us first show more clearly the notion of life to which we are leading up.

The mechanistic explanations, we said, hold good for the systems that our thought artificially detaches from the whole. But of the whole itself and of the systems which, within this whole, seem to take after it, we cannot admit *a priori* that they are mechanically explicable, for then time would be useless, and even unreal. The essence of mechanical explanation, in fact, is to regard the future and the past as calculable functions of the present, and thus to claim that *all is given.* On this hypothesis, past, present and future would be open at a glance to a superhuman intellect capable of making the calculation. Indeed, the scientists who have believed in the universality and perfect objectivity of mechanical explanations have, consciously or unconsciously, acted on a hypothesis of this kind. Laplace formulated it with the greatest precision: "An intellect which at a given instant knew all the forces with which nature is animated, and the respective situations of the beings that compose nature—supposing the said intellect were vast enough to subject these data to analysis—would embrace in the same formula the motions of the greatest bodies in the universe and those of the slightest atom: nothing would be uncertain for it, and the future, like the past, would be present to its eyes." [1] And Du Bois-Reymond: "We can imagine the knowledge of nature arrived at a point where the universal process of the world might be represented by a single mathematical formula, by one immense system of simultaneous differential equations, from which could be deduced, for each moment, the position, direction, and velocity of every atom of the world." [2] Huxley has expressed the same idea in a more concrete form: "If the fundamental proposition of evolution is true, that the entire world, living and not living, is the result of the mutual interaction, according to definite laws, of the forces possessed by the molecules of which the primitive nebulosity of the universe was composed, it is no less certain that the existing world lay, potentially, in the cosmic vapor, and that a sufficient intellect could, from a knowledge of the properties of the molecules of that vapor, have predicted, say the state of the Fauna of Great Britain in 1869, with as much certainty as one can say what will happen to the vapor of the breath in a cold winter's day." In such a doctrine, time is still spoken of: one pronounces the word, but one does not

[1] Laplace, *Introduction à la théorie analytique des probabilités* (*Œuvres complètes*, vol. vii., Paris, 1886, p. vi.).

[2] Du Bois-Reymond, *Über die Grenzen des Naturerkennens,* Leipzig, 1892.

think of the thing. For time is here deprived of efficacy and if it *does* nothing, it *is* nothing. Radical mechanism implies a metaphysic in which the totality of the real is postulated complete in eternity, and in which the apparent duration of things expresses merely the infirmity of a mind that cannot know everything at once. But duration is something very different from this for our consciousness, that is to say, for that which is most indisputable in our experience. We perceive duration as a stream against which we cannot go. It is the foundation of our being, and, as we feel, the very substance of the world in which we live. It is of no use to hold up before our eyes the dazzling prospect of a universal mathematic; we cannot sacrifice experience to the requirements of a system. That is why we reject radical mechanism.

But radical finalism is quite as unacceptable, and for the same reason. The doctrine of teleology, in its extreme form, as we find it in Leibniz for example, implies that things and beings merely realize a program previously arranged. But if there is nothing unforeseen, no invention or creation in the universe, time is useless again. As in the mechanistic hypothesis, here again it is supposed that *all is given*. Finalism thus understood is only inverted mechanism. It springs from the same postulate, with this sole difference, that in the movement of our finite intellects along successive things, whose successiveness is reduced to a mere appearance, it holds in front of us the light with which it claims to guide us, instead of putting it behind. It substitutes the attraction of the future for the impulsion of the past.

But succession remains none the less a mere appearance, as indeed does movement itself. In the doctrine of Leibniz, time is reduced to a confused perception, relative to the human standpoint, a perception which would vanish, like a rising mist, for a mind seated at the center of things.

Yet finalism is not, like mechanism, a doctrine with fixed rigid outlines. It admits of as many inflections as we like. The mechanistic philosophy is to be taken or left: it must be left if the least grain of dust, by straying from the path foreseen by mechanics, should show the slightest trace of spontaneity. The doctrine of final causes, on the contrary, will never be definitively refuted. If one form of it be put aside, it will take another. Its principle, which is essentially psychological, is very flexible. It is so extensible, and thereby so comprehensive, that one accepts something of it as soon as one rejects pure mechanism. The theory we shall put forward in this book will therefore necessarily partake of finalism to a certain extent. For that reason it is important to intimate exactly what we are going to take of it, and what we mean to leave.

Let us say at once that to thin out the Leibnizian finalism by breaking it into an infinite number of pieces seems to us a step in the wrong direction. This is, however, the tendency of the doctrine of finality. It fully realizes that if the universe as a whole is the carrying out of a plan, this cannot be demonstrated empirically, and that even of the organized world alone it is hardly easier to prove all harmonious: facts would equally well testify to the contrary. Nature sets living beings at discord with one another. She everywhere presents disorder alongside of order, retrogression alongside of progress. But, though finality cannot be affirmed either of the whole of matter or of the whole of life, might it not yet be true, says the finalist, of each organism taken separately? Is there not a wonderful division of labor, a marvelous solidarity among the parts of an organism, perfect order in infinite complexity? Does not each living being thus realize a plan immanent in its substance?—This theory consists, at bottom, in breaking up the original notion of finality into bits. It does not accept, indeed it ridicules, the idea of an *external* finality, according to which living beings are ordered with regard to each other: to suppose the grass made for the cow, the lamb for the wolf—that is all acknowledged to be absurd. But there is, we are told, an *internal* finality: each being is made for itself, all its parts conspire for the greatest good of the whole and are intelligently organized in view of that end. Such is the notion of finality which has long been classic. Finalism has shrunk to the point of never embracing more than one living being at a time. By making itself smaller, it probably thought it would offer less surface for blows.

The truth is, it lay open to them a great deal more. Radical as our own theory may appear, finality is external or it is nothing at all.

Consider the most complex and the most harmonious organism. All the elements, we are told, conspire for the greatest good of the whole. Very well, but let us not forget that each of these elements may itself be an organism in certain cases, and that in subordinating the existence of this small organism to the life of the great one we accept the principle of an *external* finality. The idea of a finality that is *always* internal is therefore a self-destructive notion. An organism is composed of tissues, each of which lives for itself. The cells of which the tissues are made have also a certain independence. Strictly speaking, if the subordination of all the elements of the individual to the individual itself were complete, we might contend that they are not organisms, reserve the name organism for the individual, and recognize only internal finality. But every one knows that these elements may possess a true autonomy. To say nothing of phagocytes, which push independence to the point of attacking the organism that nourishes them, or of germinal cells, which have their own life alongside the somatic cells—the facts of regeneration are enough: here an element or a group of elements suddenly reveals that, however limited its normal space and function, it can transcend them occasionally; it may even, in certain cases, be regarded as the equivalent of the whole.

There lies the stumbling-block of the vitalistic theories. We shall not reproach them, as is ordinarily done, with replying to the question by the question itself: the "vital principle" may indeed not explain much, but it is at least a sort of label affixed to our ignorance, so as to remind us of this occasionally,[1] while mechanism in-

[1] There are really two lines to follow in contemporary neo-vitalism: on the one hand, the assertion that pure mechanism is insufficient, which assumes great authority when made by such scientists as Driesch or Reinke, for example; and, on the other hand, the hypotheses which this vitalism superposes on mechanism (the "entelechies" of Driesch, and the "dominants" of Reinke, etc.). Of these two parts, the former is perhaps the more interesting. See the admirable studies of Driesch—*Die Lokalisation morphogenetischer Vorgänge*, Leipzig, 1899; *Die organischen Regulationen*, Leipzig, 1901; *Naturbegriffe und Natururteile*, Leipzig, 1904; *Der Vitalismus als Geschichte und als Lehre*, Leipzig,

vites us to ignore that ignorance. But the position of vitalism is rendered very difficult by the fact that, in nature, there is neither purely internal finality nor absolutely distinct individuality. The organized elements composing the individual have themselves a certain individuality, and each will claim its vital principle if the individual pretends to have its own. But, on the other hand, the individual itself is not sufficiently independent, not sufficiently cut off from other things, for us to allow it a "vital principle" of its own. An organism such as a higher vertebrate is the most individuated of all organisms; yet, if we take into account that it is only the development of an ovum forming part of the body of its mother and of a spermatozoon belonging to the body of its father, that the egg (*i.e.* the ovum fertilized) is a connecting link between the two progenitors since it is common to their two substances, we shall realize that every individual organism, even that of a man, is merely a bud that has sprouted on the combined body of both its parents. Where, then, does the vital principle of the individual begin or end? Gradually we shall be carried further and further back, up to the individual's remotest ancestors: we shall find him solidary with each of them, solidary with that little mass of protoplasmic jelly which is probably at the root of the genealogical tree of life. Being, to a certain extent, one with this primitive ancestor, he is also solidary with all that descends from the ancestor in divergent directions. In this sense each individual may be said to remain united

1905; and of Reinke—*Die Welt als Tat*, Berlin, 1899; *Einleitung in die theoretische Biologie*, Berlin, 1901; *Philosophie der Botanik*, Leipzig, 1905.

with the totality of living beings by invisible bonds. So it is of no use to try to restrict finality to the individuality of the living being. If there is finality in the world of life, it includes the whole of life in a single indivisible embrace. This life common to all the living undoubtedly presents many gaps and incoherences, and again it is not so mathematically *one* that it cannot allow each being to become individualized to a certain degree. But it forms a single whole, none the less; and we have to choose between the out-and-out negation of finality and the hypothesis which co-ordinates not only the parts of an organism with the organism itself, but also each living being with the collective whole of all others.

Finality will not go down any easier for being taken as a powder. Either the hypothesis of a finality immanent in life should be rejected as a whole, or it must undergo a treatment very different from pulverization.

The error of radical finalism, as also that of radical mechanism, is to extend too far the application of certain concepts that are natural to our intellect. Originally, we think only in order to act. Our intellect has been cast in the mold of action. Speculation is a luxury, while action is a necessity. Now, in order to act, we begin by proposing an end; we make a plan, then we go on to the detail of the mechanism which will bring it to pass. This latter operation is possible only if we know what we can reckon on. We must therefore have managed to extract resemblances from nature, which enable us to anticipate the future. Thus we must, consciously or unconsciously, have made use of the law of causality. Moreover, the more sharply the idea of efficient causality is defined in our mind, the more it takes the form of a *mechanical* causality. And this scheme, in its turn, is the more mathematical according as it expresses a more rigorous necessity. That is why we have only to follow the bent of our mind to become mathematicians. But, on the other hand, this natural mathematics is only the rigid unconscious skeleton beneath our conscious supple habit of linking the same causes to the same effects; and the usual object of this habit is to guide actions inspired by intentions, or, what comes to the same, to direct movements combined with a view to reproducing a pattern. We are born artisans as we are born geometricians, and indeed we are geometricians only because we are artisans. Thus the human intellect, inasmuch as it is fashioned for the needs of human action, is an intellect which proceeds at the same time by intention and by calculation, by adapting means to ends and by thinking out mechanisms of more and more geometrical form. Whether nature be conceived as an immense machine regulated by mathematical laws, or as the realization of a plan, these two ways of regarding it are only the consummation of two tendencies of mind which are complementary to each other, and which have their origin in the same vital necessities.

For that reason, radical finalism is very near radical mechanism on many points. Both doctrines are reluctant to see in the course of things generally, or even simply in the development of life, an unforeseeable creation of form. In considering reality, mechanism regards only the aspect of similarity or repetition. It is therefore dominated by this law, that in nature there is only *like* reproducing *like*. The more the geometry in mechanism is emphasized, the less can mechanism admit that anything is ever created, even pure form. In so far as we are geometricians, then, we reject the unforeseeable. We might accept it, assuredly, in so far as we are artists, for art lives on creation and implies a latent belief in the spontaneity of nature. But disinterested art is a luxury, like pure speculation. Long before being artists, we are artisans; and all fabrication, however rudimentary, lives on likeness and repetition, like the natural geometry which serves as its fulcrum. Fabrication works on models which it sets out to reproduce; and even when it invents, it proceeds, or imagines itself to proceed, by a new arrangement of elements already known. Its principle is that "we must have like to produce like." In short, the strict application of the principle of finality, like that of the principle of mechanical causality, leads to the conclusion that "all is given." Both principles say the same thing in their respective languages, because they respond to the same need.

That is why again they agree in doing away with time. Real duration is that duration which gnaws on things, and leaves on them the mark of its tooth. If everything is in time, everything changes inwardly, and the same concrete reality never recurs. Repetition is therefore possible only in the abstract: what is repeated is some aspect that our senses, and especially our intellect, have singled out from reality, just because our action, upon which all the effort of our intellect is directed, can move only among repetitions. Thus, concentrated on that which repeats, solely preoccupied in welding the same to the same, intellect turns away from the vision of time. It dislikes what is fluid, and solidifies everything it touches. We do not *think* real time. But we *live* it, because life transcends intellect. The feeling we have of our evolution and of the evolution of all things in pure duration is there, forming around the intellectual concept properly so-called an indistinct fringe that fades off into darkness. Mechanism and finalism agree in taking account only of the bright nucleus shining in the center. They forget that this nucleus has been formed out of the rest by condensation, and that the whole must be used, the fluid as well as and more than the condensed, in order to grasp the inner movement of life.

Indeed, if the fringe exists, however delicate and indistinct, it should have more importance for philosophy than the bright nucleus it surrounds. For it is its presence that enables us to affirm that the nucleus is a nucleus, that pure intellect is a contraction, by condensation, of a more extensive power. And, just because this vague intuition is of no help in directing our action on things, which action takes place exclusively on the surface of reality, we may presume that it is to be exercised not merely on the surface, but below.

As soon as we go out of the encasings in which radical mechanism and radical finalism confine our thought, reality appears as a ceaseless upspringing of something new, which has no sooner arisen to make the present than it has already fallen back into the past; at this exact moment it falls under the glance of the intellect, whose eyes are ever turned to the rear. This is already the case with our inner life. For each of our acts we shall easily find antecedents of which it may in some sort be said to be the mechanical resultant. And it may equally well be said that each action is the realization of an intention. In this sense mechanism is everywhere, and finality everywhere, in the evolution of our conduct. But if our action be one that involves the whole of our person and is truly ours, it could not have been foreseen, even though its antecedents explain it when once it has been accomplished. And though it be the realizing of an intention, it differs, as a present and *new* reality, from the intention, which can never aim at anything but recommencing or rearranging the past. Mechanism and finalism are therefore, here, only external views of our conduct. They extract its intellectuality. But our conduct slips between them and extends much further. Once again, this does not mean that free action is capricious, unreasonable action. To behave according to caprice is to oscillate mechanically between two or more ready-made alternatives and at length to settle on one of them; it is no real maturing of an internal state, no real evolution; it is merely—however paradoxical the assertion may seem—bending the will to imitate the mechanism of the intellect. A conduct that is truly our own, on the contrary, is that of a will which does not try to counterfeit intellect, and which, remaining itself—that is to say, evolving—ripens gradually into acts which the intellect will be able to resolve indefinitely into intelligible elements without ever reaching its goal. The free act is incommensurable with the idea, and its "rationality" must be defined by this very incommensurability, which admits the discovery of as much intelligibility within it as we will. Such is the character of our own evolution; and such also, without doubt, that of the evolution of life.

Our reason, incorrigibly presumptuous, imagines itself possessed, by right of birth or by right of conquest, innate or acquired, of all the essential elements of the knowledge of truth. Even where it confesses that it does not know the object presented to it, it believes that its ignorance consists only in not knowing which one of its time-honored categories suits the new object. In what drawer, ready to open, shall we put it? In what garment, already cut out, shall we clothe it? Is it this, or that, or the other thing? And "this," and "that," and "the other thing" are always something already conceived, already known. The idea that for a new object we might have to create a new concept, perhaps a new method of thinking, is deeply repugnant to us. The history of philosophy is there, however, and shows us the eternal conflict of systems, the impossibility of satisfactorily getting the real into the ready-made garments of our ready-made concepts, the necessity of making to measure. But, rather than go to this extremity, our reason prefers to announce once for all, with a proud modesty, that it has to do only with the relative, and that the absolute is not in its province. This preliminary declaration enables it to apply its habitual method of thought without any scruple, and thus, under pretense that it does not touch the absolute, to make absolute judgments upon everything. Plato was the first to set up the theory that to know the real consists in finding its Idea, that is to say, in forcing it into a pre-existing frame already at our disposal—as if we implicitly possessed universal knowledge. But this belief is natural to the human intellect, always engaged as it is in determining under what former heading it shall catalogue any new object; and it may be said that, in a certain sense, we are all born Platonists.

Nowhere is the inadequacy of this method so obvious as in theories of life. If, in evolving in the direction of the vertebrates in general, of man and intellect in particular, life has had to abandon by the way many elements incompatible with this particular mode of organization and consign them, as we shall show, to other lines of development, it is the totality of these elements that we must find again and rejoin to the intellect proper, in order to grasp the true nature of vital activity. And we shall probably be aided in this by the fringe of vague intuition that surrounds our distinct—that is, intellectual—representation. For what can this useless fringe be, if not that part of the evolving principle which has not shrunk to the peculiar form of our organization, but has settled around it unasked for, unwanted? It is there, accordingly, that we must look for hints to expand the intellectual form of our thought; from there shall we derive the impetus necessary to lift us above ourselves. To form an idea of the whole of life cannot consist in combining simple ideas that have been left behind in us by life itself in the course of its evolution. How could the part be equivalent to the whole, the content to the container, a by-product of the vital operation to the operation itself? Such, however, is our illusion when we define the evolution of life as a "passage from the homogeneous to the heterogeneous," or by any other concept obtained by putting fragments of intellect side by side. We place ourselves in one of the points where evolution comes to a head—the principal one, no doubt, but not the only one; and there we do not even take all we find, for of the intellect we keep only one or two of the concepts by which it expresses itself; and it is this part of a part that we declare representative of the whole, of something indeed which goes beyond the concrete whole, I mean of the evolution movement of which this "whole" is only the present stage! The truth is, that to represent this the entire intellect would not be too much—nay, it would not be enough. It would be necessary to add to it what we find in every other terminal point of evolution. And these diverse and divergent elements must be considered as so many extracts which are, or at least which were, in their humblest form, mutually complementary. Only then might we have an inkling of the real nature of the evolution movement; and even then we should fail to grasp it completely, for we should still be dealing only with the evolved, which is a result, and not with evolution itself, which is the act by which the result is obtained.

Such is the philosophy of life to which we are leading up. It claims to transcend both mechanism and finalism; but, as we announced at the beginning, it is nearer the second doctrine than the first. It will not be amiss to dwell on this point, and show more precisely how far this philosophy of life resembles finalism and wherein it is different.

Like radical finalism, although in a vaguer form, our philosophy represents the organized world as a harmonious whole. But this harmony is far from being as perfect as it has been claimed to be. It admits of much discord, because each species, each individual even, retains only a certain impetus from the universal vital impulsion and tends to use this energy in its own interest. In this consists *adaptation*. The species and the individual thus think only of themselves—whence arises a possible conflict with other forms of life. Harmony, therefore, does not exist in fact; it exists rather in principle; I mean that the original impetus is a *common* impetus, and the higher we ascend the stream of life the more do diverse tendencies appear complementary to each other. Thus the wind at a street corner divides into diverging currents which are all one and the same gust. Harmony, or rather "complementarity," is revealed only in the mass, in tendencies rather than in states. Especially (and this is the point on which finalism has been most seriously mistaken) harmony is rather behind us than before. It is due to an identity of impulsion and not to a common aspiration. It would be futile to try to assign to life an end, in the human sense of the word. To speak of an end is to think of a preexisting model which has only to be realized. It is to suppose, therefore, that all is given, and that the future can be read in the present. It is to believe that life, in its movement and in its entirety, goes to work like our intellect, which is only a motionless and fragmentary view of life, and which naturally takes its stand outside of time. Life, on the contrary, progresses and *endures* in time. Of course, when once the road has been traveled, we can glance over it, mark its direction, note this in psychological terms and speak as if there had been pursuit of an end. Thus shall we speak ourselves. But, of the road which was going to be traveled, the human mind could have nothing to say, for the road has been created *pari passu* with the act of traveling over it, being nothing but the direction of this act itself. At every instant, then, evolution must admit of a psychological interpretation which is, from our point of view, the best interpretation; but this explanation has neither value nor even significance except retrospectively. Never could the finalistic interpretation, such as we shall propose it, be taken for an anticipation of the future. It is a particular mode of viewing the past in the light of the present. In short, the classic conception of finality postulates at once too much and too little: it is both too wide and too narrow. In explaining life by intellect, it limits too much the meaning of life: intellect, such at least as we find it in ourselves, has been fashioned by evolution during the course of progress; it is cut out of something larger, or, rather, it is only the projection, necessarily on a plane, of a reality that possesses both relief and depth. It is this more comprehensive reality that true finalism ought to reconstruct, or, rather, if possible, embrace in one view. But, on the other hand, just because it goes beyond intellect—the faculty of connecting the same with the same, of perceiving and also of producing repetitions—this reality is undoubtedly creative, *i.e.* productive of effects in which it expands and transcends its own being. These effects were therefore not given in it in advance, and so it could not take them for ends, although, when once produced, they admit of a rational interpretation, like that of the manufactured article that has reproduced a model. In short, the theory of final causes does not go far enough when it confines itself to ascribing some intelligence to nature, and it goes too far when it supposes a pre-existence of the future in the present in the form of idea. And the second theory, which sins by excess, is the outcome of the first, which sins by defect. In place of intellect proper must be substituted the more comprehensive reality of which intellect is only the contraction. The future then appears as expanding the present: it was not, therefore, contained in the present in the form of a represented end. And yet, once realized, it will explain the present as much as the present explains it, and even more; it must be viewed as an end as much as, and more than, a result. Our intellect has a right to consider the future abstractly from its habitual point of view, being itself an abstract view of the cause of its own being.

It is true that the cause may then seem beyond our grasp. Already the finalist theory of life eludes all precise verification. What if we go beyond it in one of its directions? Here, in fact, after a necessary digression, we are back at the question which we regard as essential: can the insufficiency of mechanism be proved by facts? We said that if this demonstration is possible, it is on condition of frankly accepting the evolutionist hypothesis. We must now show that if mechanism is insufficient to account for evolution, the way of proving this insufficiency is not to stop at the classic conception of finality, still less to contract or attenuate it, but, on the contrary, to go further.

Let us indicate at once the principle of our demonstration. We said of life that, from its origin, it is the continuation of one and the same impetus, divided into divergent lines of evolution. Something has grown, something has developed by a series of additions which have been so many creations. This very development has brought about a dissociation of tendencies which were unable to grow beyond a certain point without becoming mutually incompatible. Strictly speaking, there is nothing to prevent our imagining that the evolution of life might have taken place in one single individual by means of a series of transformations spread over thousands of ages. Or, instead of a single individual, any number might be supposed, succeeding each other in a unilinear series. In both cases evolution would have had, so to speak, one dimension only. But evolution has actually taken place through millions of individuals, on divergent lines, each ending at a crossing from which new paths radiate, and so on indefinitely. If our hypothesis is justified, if the essential causes working along these diverse roads are of psychological nature, they must keep something in common in spite of the divergence of their effects, as school-fellows long separated keep the same memories of boyhood. Roads may fork or by-ways be opened along which dissociated elements may evolve in an independent manner, but nevertheless it is in virtue of the primitive impetus of the whole that the movement of the parts continues. Something of the whole, therefore, must abide in the parts; and this common element will be evident to us in some way, perhaps by the presence of identical organs in very different organisms. Suppose, for an instant, that the mechanistic explanation is the true one: evolution must then have occurred through a series of accidents added to one another, each new accident being preserved by selection if it is advantageous to that sum of former advantageous accidents which the present form of the living being represents. What likelihood is there that, by two entirely different series of accidents being added together, two entirely different evolutions will arrive at similar results? The more two lines of evolution diverge, the less probability is there that accidental outer influences or accidental inner variations bring about the construction of the same apparatus upon them, especially if there was no trace of this apparatus at the moment of divergence. But such similarity of the two products would be natural, on the contrary, in a hypothesis like ours: even in the latest channel there would be something of the impulsion received at the source. *Pure mechanism, then, would be refutable, and finality, in the special sense in which we understand it, would be demonstrable in a certain aspect, if it could be proved that life may manufacture the like apparatus, by unlike means, on divergent lines of evolution; and the strength of the proof would be proportional both to the divergency between the lines of evolution thus chosen and to the complexity of the similar structures found in them.*

It will be said that resemblance of structure is due to sameness of the general conditions in which life has evolved, and that these permanent outer conditions may have imposed the same direction on the forces constructing this or that apparatus, in spite of the diversity of transient outer influences and accidental inner changes. We are not, of course, blind to the rôle which the concept of *adaptation* plays in the science of today. Biologists certainly do not all make the same use of it. Some think the outer conditions capable of causing change in organisms in a *direct* manner, in a definite direction, through physico-chemical alterations induced by them in the living substance; such is the hypothesis of Eimer, for example. Others, more faithful to the spirit of Darwinism, believe the influence of conditions works *indirectly* only, through favoring, in the struggle for life, those representatives of a species which the chance of birth has best adapted to the environment. In other words, some attribute a *positive* influence to outer conditions, and say that they actually *give rise to* variations, while the others say these conditions have only a *negative* influence and merely *eliminate* variations. But, in both cases, the outer conditions are supposed to bring about a precise adjustment of the organism to its circumstances. Both parties, then, will attempt to explain mechanically, by adaptation to similar conditions, the similarities of structure which we think are the strongest argument against mechanism. So we must at once indicate in a general way, before passing to the detail, why explanations from "adaptation" seem to us insufficient.

Let us first remark that, of the two hypotheses just described, the latter is the only one which is not equivocal. The Darwinian idea of adaptation by automatic elimination of the unadapted is a simple and clear idea. But, just because it attributes to the outer cause which controls evolution a merely negative influence, it has great difficulty in accounting for the progressive and, so to say, rectilinear development of complex apparatus such as we are about to examine. How much greater will this difficulty be in the case of the similar structure of two extremely complex organs on two entirely different lines of evolution! An accidental variation, however minute, implies the working of a great number of small physical and chemical causes. An accumulation of accidental variations, such as would be necessary to produce a complex structure, requires therefore the concurrence of an almost infinite number of infinitesimal causes. Why should these causes, entirely accidental, recur the same, and in the same order, at different points of space and time? No one will hold that this is the case, and the Darwinian himself will probably merely maintain that identical effects may arise from different causes, that more than one road leads to the same spot. But let us not be fooled by a metaphor. The place reached does not give the form of the road that leads there; while an organic structure is just the accumulation of those small differences which evolution has had to go through in order to achieve it. The struggle for life and natural selection can be of no use to us in solving this part of the problem, for we are not concerned here with what has perished, we have to do only with what has survived. Now, we see that identical structures have been formed on independent lines of evolution by a gradual accumulation of effects. How can accidental causes, occurring in an accidental order, be supposed to have repeatedly come to the same result, the causes being infinitely numerous and the effect infinitely complicated?

The principle of mechanism is that "the same causes produce the same effects." This principle, of course, does not always imply that the same effects must have the same causes; but it does involve this consequence in the particular case in which the causes remain visible in the effect that they produce and are indeed its constitutive elements. That two walkers starting from different points and wandering at random should finally meet, is no great wonder. But that, throughout their walk, they should describe two identical curves exactly superposable on each other, is altogether unlikely. The improbability will be the greater, the more complicated the routes; and it will become impossibility, if the zigzags are infinitely complicated. Now, what is this complexity of zigzags as compared with that of an organ in which thousands of different cells, each being itself a kind of organism, are arranged in a definite order?

Let us turn, then, to the other hypothesis, and see how it would solve the problem. Adaptation, it says, is not merely elimination of the unadapted; it is due to the positive influence of outer conditions that have molded the organism on their own form. This time, similarity of effects will be explained by similarity of cause. We shall remain, apparently, in pure mechanism. But if we look closely, we shall see that the explanation is merely verbal, that we are again the dupes of words, and that the trick of the solution consists in taking the term "adaptation" in two entirely different senses at the same time.

If I pour into the same glass, by turns, water and wine, the two liquids will take the same form, and the sameness in form will be due to the sameness in adaptation of content to container. Adaptation, here, really means mechanical adjustment. The reason is that the form to which the matter has adapted itself was there, ready-made, and has forced its own shape on the matter. But, in the adaptation of an organism to the circumstances it has to live in, where is the pre-existing form awaiting its matter? The circumstances are not a mold into which life is inserted and whose form life adopts: this is indeed to be fooled by a metaphor. There is no form yet, and the life must create a form for itself, suited to the circumstances which are made for it. It will have to make the best of these circumstances, neutralize their inconveniences and utilize their advantages—in short, respond to outer actions by building up a machine which has no resemblance to them. Such adapting is not *repeating*, but *replying*,—an entirely different thing. If there is still adaptation, it will be in the sense in which one may say of the solution of a problem of geometry, for example, that it is adapted to the conditions. I grant indeed that adaptation so understood explains why different evolutionary processes result in similar forms: the same problem, of course, calls for the same solution. But it is necessary then to introduce, as for the solution of a problem of geometry, an intelligent activity, or at least a cause which behaves in the same way. This is to bring in finality again, and a finality this time more than ever charged with anthropomorphic elements. In a word, if the adaptation is passive, if it is mere repetition in the relief of what the conditions give in the mold, it will build up nothing that one tries to make it build; and if it is active, capable of responding by a calculated solution to the problem which is set out in the conditions, that is going further than we do—too far, indeed, in our opinion—in the direction we indicated in the beginning. But the truth is that there is a surreptitious passing from one of these two meanings to the other, a flight for refuge to the first whenever one is about to be caught *in flagrante delicto* of finalism by employing the second. It is really the second which serves the usual practice of science, but it is the first that generally provides its philosophy. In any *particular* case one talks as if the process of adaptation were an effort of the organism to build up a machine capable of turning external circumstances to the best possible account: then one speaks of adaptation *in general* as if it were the very impress of circumstances, passively received by an indifferent matter.

But let us come to the examples. It would be interesting first to institute here a general comparison between plants and animals. One cannot fail to be struck with the parallel progress which has been accomplished, on both sides, in the direction of sexuality. Not only is fecundation itself the same in higher plants and in animals, since it consists, in both, in the union of two nuclei that differ in their properties and structure before their union and immediately after become equivalent to each other; but the preparation of sexual elements goes on in both under like conditions: it consists essentially in the reduction of the number of chromosomes and the rejection of a certain quantity of chromatic substance.[1] Yet vegetables and animals have evolved on independent lines, favored by unlike circumstances, opposed by unlike obstacles. Here are two great series which have gone on diverging. On either line, thousands and thousands of causes have combined to determine the morphological and functional evolution. Yet these infinitely complicated causes have been consummated, in each series, in the same effect. And this effect could hardly be called a phenomenon of "adaptation": where is the adaptation, where is the pressure of external circumstances? There is no striking utility in sexual generation; it has been interpreted in the most diverse ways; and some very acute enquirers even

[1] P. Guérin, *Les Connaissances actuelles sur la fécondation chez les phanérogames*, Paris, 1904, pp. 144-148. Cf. Delage, *L'Hérédité*, 2nd edition, 1903, pp. 140 ff.

regard the sexuality of the plant, at least, as a luxury which nature might have dispensed with.[1] But we do not wish to dwell on facts so disputed. The ambiguity of the term "adaptation," and the necessity of transcending both the point of view of mechanical causality and that of anthropomorphic finality, will stand out more clearly with simpler examples. At all times the doctrine of finality has laid much stress on the marvelous structure of the sense-organs, in order to liken the work of nature to that of an intelligent workman. Now, since these organs are found, in a rudimentary state, in the lower animals, and since nature offers us many intermediaries between the pigment-spot of the simplest organisms and the infinitely complex eye of the vertebrates, it may just as well be alleged that the result has been brought about by natural selection perfecting the organ automatically. In short, if there is a case in which it seems justifiable to invoke adaptation, it is this particular one. For there may be discussion about the function and meaning of such a thing as sexual generation, in so far as it is related to the conditions in which it occurs; but the relation of the eye to light is obvious, and when we call this relation an adaptation, we must know what we mean. If, then, we can show, in this privileged case, the insufficiency of the principles invoked on both sides, our demonstration will at once have reached a high degree of generality.

Let us consider the example on which the advocates of finality have always insisted: the structure of such an organ as the human eye. They have had no difficulty in

[1] Möbius, *Beiträge zur Lehre von der Fortpflanzung der Gewächse*, Jena, 1897, pp. 203-206 in particular. Cf. Hartog, "Sur les phénomènes de reproduction" (*Année biologique*, 1895, pp. 707-709).

showing that in this extremely complicated apparatus all the elements are marvelously co-ordinated. In order that vision shall operate, says the author of a well-known book on *Final Causes,* "the sclerotic membrane must become transparent in one point of its surface, so as to enable luminous rays to pierce it . . . ; the cornea must correspond exactly with the opening of the socket . . . ; behind this transparent opening there must be refracting media . . . ; there must be a retina[1] at the extremity of the dark chamber . . . ; perpendicular to the retina there must be an innumerable quantity of transparent cones permitting only the light directed in the line of their axes to reach the nervous membrane,"[2] etc., etc. In reply, the advocate of final causes has been invited to assume the evolutionist hypothesis. Everything is marvelous, indeed, if one consider an eye like ours, in which thousands of elements are co-ordinated in a single function. But take the function at its origin, in the Infusorian, where it is reduced to the mere impressionability (almost purely chemical) of a pigment-spot to light: this function, possibly only an accidental fact in the beginning, may have brought about a slight complication of the organ, which again induced an improvement of the function. It may have done this either directly, through some unknown mechanism, or indirectly, merely through the effect of the advantages it brought to the living being and the hold it thus offered to natural selection. Thus the progressive formation of an eye as well contrived as ours would be explained by an almost infinite number of actions and reactions between the function

[1] Paul Janet, *Les Causes finales*, Paris, 1876, p. 83.

[2] *Ibid.* p. 80.

and the organ, without the intervention of other than mechanical causes.

The question is hard to decide, indeed, when put directly between the function and the organ, as is done in the doctrine of finality, as also mechanism itself does. For organ and function are terms of different nature, and each conditions the other so closely that it is impossible to say *a priori* whether in expressing their relation we should begin with the first, as does mechanism, or with the second, as finalism requires. But the discussion would take an entirely different turn, we think, if we began by comparing together two terms of the same nature, an organ with an organ, instead of an organ with its function. In this case, it would be possible to proceed little by little to a solution more and more plausible, and there would be the more chance of a successful issue the more resolutely we assumed the evolutionist hypothesis.

Let us place side by side the eye of a vertebrate and that of a mollusk such as the common Pecten. We find the same essential parts in each, composed of analogous elements. The eye of the Pecten presents a retina, a cornea, a lens of cellular structure like our own. There is even that peculiar inversion of retinal elements which is not met with, in general, in the retina of the invertebrates. Now, the origin of mollusks may be a debated question, but, whatever opinion we hold, all are agreed that mollusks and vertebrates separated from their common parent-stem long before the appearance of an eye so complex as that of the Pecten. Whence, then, the structural analogy?

Let us question on this point the two opposed systems of evolutionist explanation in turn—the hypothesis of purely accidental variations, and that of a variation directed in a definite way under the influence of external conditions.

The first, as is well known, is presented today in two quite different forms. Darwin spoke of very slight variations being accumulated by natural selection. He was not ignorant of the facts of sudden variation; but he thought these "sports," as he called them, were only monstrosities incapable of perpetuating themselves; and he accounted for the genesis of species by an accumulation of *insensible* variations.[1] Such is still the opinion of many naturalists. It is tending, however, to give way to the opposite idea that a new species comes into being all at once by the simultaneous appearance of several new characters, all somewhat different from the previous ones. This latter hypothesis, already proposed by various authors, notably by Bateson in a remarkable book,[2] has become deeply significant and acquired great force since the striking experiments of Hugo de Vries. This botanist, working on the *Œnothera Lamarckiana,* obtained at the end of a few generations a certain number of new species. The theory he deduces from his experiments is of the highest interest. Species pass through alternate periods of stability and transformation. When the period of "mutability" occurs, unexpected forms spring forth in a great number of different directions.[3]—We will not attempt to take sides between this hypoth-

[1] Darwin, *Origin of Species*, chap. ii.

[2] Bateson, *Materials for the Study of Variation*, London, 1894, especially pp. 567 ff. Cf. Scott, "Variations and Mutations" (*American Journal of Science*, Nov. 1894).

[3] De Vries, *Die Mutationstheorie*, Leipzig, 1901-1903. Cf., by the same author, *Species and Varieties*, Chicago, 1905.

esis and that of insensible variations. Indeed, perhaps both are partly true. We wish merely to point out that if the variations invoked are accidental, they do not, whether small or great, account for a similarity of structure such as we have cited.

Let us assume, to begin with, the Darwinian theory of insensible variations, and suppose the occurrence of small differences due to chance, and continually accumulating. It must not be forgotten that all the parts of an organism are necessarily co-ordinated. Whether the function be the effect of the organ or its cause, it matters little; one point is certain—the organ will be of no use and will not give selection a hold unless it functions. However the minute structure of the retina may develop, and however complicated it may become, such progress, instead of favoring vision, will probably hinder it if the visual centers do not develop at the same time, as well as several parts of the visual organ itself. If the variations are accidental, how can they ever agree to arise in every part of the organ at the same time, in such way that the organ will continue to perform its function? Darwin quite understood this; it is one of the reasons why he regarded variation as insensible.[1] For a difference which arises accidentally at one point of the visual apparatus, if it be very slight, will not hinder the functioning of the organ; and hence this first accidental variation can, in a sense, *wait for* complementary variations to accumulate and raise vision to a higher degree of perfection. Granted; but while the insensible variation does not hinder the functioning of the eye, neither does it help it, so long as the variations that are complementary do not

[1] Darwin, *Origin of Species*, chap. vi.

occur. How, in that case, can the variation be retained by natural selection? Unwittingly one will reason as if the slight variation were a toothing stone set up by the organism and reserved for a later construction. This hypothesis, so little conformable to the Darwinian principle, is difficult enough to avoid even in the case of an organ which has been developed along one single main line of evolution, *e.g.* the vertebrate eye. But it is absolutely forced upon us when we observe the likeness of structure of the vertebrate eye and that of the mollusks. How could the same small variations, incalculable in number, have ever occurred in the same order on two independent lines of evolution, if they were purely accidental? And how could they have been preserved by selection and accumulated in both cases, the same in the same order, when each of them, taken separately, was of no use?

Let us turn, then, to the hypothesis of sudden variations, and see whether it will solve the problem. It certainly lessens the difficulty on one point, but it makes it much worse on another. If the eye of the mollusk and that of the vertebrate have both been raised to their present form by a relatively small number of sudden leaps, I have less difficulty in understanding the resemblance of the two organs than if this resemblance were due to an incalculable number of infinitesimal resemblances acquired successively: in both cases it is chance that operates, but in the first case chance is not required to work the miracle it would have to perform in the second. Not only is the number of resemblances to be added somewhat reduced, but I can also understand better how each could be preserved and added to the others; for the elementary variation is now considerable enough to be an advantage to the living being, and so to lend itself to the play of selection. But here there arises another problem, no less formidable, viz., how do all the parts of the visual apparatus, suddenly changed, remain so well co-ordinated that the eye continues to exercise its function? For the change of one part alone will make vision impossible, unless this change is absolutely infinitesimal. The parts must then all change at once, each consulting the others. I agree that a great number of unco-ordinated variations may indeed have arisen in less fortunate individuals, that natural selection may have eliminated these, and that only the combination fit to endure, capable of preserving and improving vision, has survived. Still, this combination had to be produced. And, supposing chance to have granted this favor once, can we admit that it repeats the self-same favor in the course of the history of a species, so as to give rise, every time, all at once, to new complications marvelously regulated with reference to each other, and so related to former complications as to go further on in the same direction? How, especially, can we suppose that by a series of mere "accidents" these sudden variations occur, the same, in the same order—involving in each case a perfect harmony of elements more and more numerous and complex—along two independent lines of evolution?

The law of correlation will be invoked, of course; Darwin himself appealed to it.[1] It will be alleged that a change is not localized in a single point of the organism, but has its necessary recoil on other points. The examples cited by Darwin remain classic: white cats with blue

[1] Darwin, *Origin of Species*, chap. i.

eyes are generally deaf; hairless dogs have imperfect dentition, etc.—Granted; but let us not play now on the word "correlation." A collective whole of *solidary* changes is one thing, a system of *complementary* changes—changes so co-ordinated as to keep up and even improve the functioning of an organ under more complicated conditions—is another. That an anomaly of the pilous system should be accompanied by an anomaly of dentition is quite conceivable without our having to call for a special principle of explanation; for hair and teeth are similar formations,[1] and the same chemical change of the germ that hinders the formation of hair would probably obstruct that of teeth: it may be for the same sort of reason that white cats with blue eyes are deaf. In these different examples the "correlative" changes are only *solidary* changes (not to mention the fact that they are really *lesions*, namely, diminutions or suppressions, and not additions, which makes a great difference). But when we speak of "correlative" changes occurring suddenly in the different parts of the eye, we use the word in an entirely new sense: this time there is a whole set of changes not only simultaneous, not only bound together by community of origin, but so co-ordinated that the organ keeps on performing the same simple function, and even performs it better. That a change in the germ, which influences the formation of the retina, may affect at the same time also the formation of the cornea, the iris, the lens, the visual centers, etc., I admit, if necessary, although they are formations that differ much more

[1] On this homology of hair and teeth, see Brandt, "Über . . . eine mutmassliche Homologie der Haare und Zahne" (*Biol. Centralblatt*, vol. xviii., 1898, especially pp. 262 ff.).

from one another in their original nature than do probably hair and teeth. But that all these simultaneous changes should occur in such a way as to improve or even merely maintain vision, this is what, in the hypothesis of sudden variation, I cannot admit, unless a mysterious principle is to come in, whose duty it is to watch over the interest of the function. But this would be to give up the idea of "accidental" variation. In reality, these two senses of the word "correlation" are often interchanged in the mind of the biologist, just like the two senses of the word "adaptation." And the confusion is almost legitimate in botany, that science in which the theory of the formation of species by sudden variation rests on the firmest experimental basis. In vegetables, function is far less narrowly bound to form than in animals. Even profound morphological differences, such as a change in the form of leaves, have no appreciable influence on the exercise of function, and so do not require a whole system of complementary changes for the plant to remain fit to survive. But it is not so in the animal, especially in the case of an organ like the eye, a very complex structure and very delicate function. Here it is impossible to identify changes that are simply solidary with changes which are also complementary. The two senses of the word "correlation" must be carefully distinguished; it would be a downright paralogism to adopt one of them in the premises of the reasoning, and the other in the conclusion. And this is just what is done when the principle of correlation is invoked in explanations of *detail* in order to account for complementary variations, and then correlation *in general* is spoken of as if it were any group of variations provoked by any variation of the germ. Thus, the notion of correlation is first used in current science as it might be used by an advocate of finality; it is understood that this is only a convenient way of expressing oneself, that one will correct it and fall back on pure mechanism when explaining the nature of the principles and turning from science to philosophy. And one does then come back to pure mechanism, but only by giving a new meaning to the word "correlation"—a meaning which would now make correlation inapplicable to the detail it is called upon to explain.

To sum up, if the accidental variations that bring about evolution are insensible variations, some good genius must be appealed to—the genius of the future species—in order to preserve and accumulate these variations, for selection will not look after this. If, on the other hand, the accidental variations are sudden, then, for the previous function to go on or for a new function to take its place, all the changes that have happened together must be complementary. So we have to fall back on the good genius again, this time to obtain the *convergence* of *simultaneous* changes, as before to be assured of the *continuity of direction* of *successive* variations. But in neither case can parallel development of the same complex structures on independent lines of evolution be due to a mere accumulation of accidental variations. So we come to the second of the two great hypotheses we have to examine. Suppose the variations are due, not to accidental and inner causes, but to the direct influence of outer circumstances. Let us see what line we should have to take, on this hypothesis, to account for the resemblance of eye-structure in two series that are independent of each other from the phylogenetic point of view.

Though mollusks and vertebrates have evolved separately, both have remained exposed to the influence of light. And light is a physical cause bringing forth certain definite effects. Acting in a continuous way, it has been able to produce a continuous variation in a constant direction. Of course it is unlikely that the eye of the vertebrate and that of the mollusk have been built up by a series of variations due to simple chance. Admitting even that light enters into the case as an instrument of selection, in order to allow only useful variations to persist, there is no possibility that the play of chance, even thus supervised from without, should bring about in both cases the same juxtaposition of elements co-ordinated in the same way. But it would be different supposing that light acted directly on the organized matter so as to change its structure and somehow adapt this structure to its own form. The resemblance of the two effects would then be explained by the identity of the cause. The more and more complex eye would be something like the deeper and deeper imprint of light on a matter which, being organized, possesses a special aptitude for receiving it.

But can an organic structure be likened to an imprint? We have already called attention to the ambiguity of the term "adaptation." The gradual complication of a form which is being better and better adapted to the mold of outward circumstances is one thing, the increasingly complex structure of an instrument which derives more and more advantage from these circumstances is another. In the former case, the matter merely receives an imprint; in the second, it reacts positively, it solves a problem. Obviously it is this second sense of the word "adapt" that is used when one says that the eye has become better and better adapted to the influence of light. But one passes more or less unconsciously from this sense to the other, and a purely mechanistic biology will strive to make the *passive* adaptation of an inert matter, which submits to the influence of its environment, mean the same as the *active* adaptation of an organism which derives from this influence an advantage it can appropriate. It must be owned, indeed, that Nature herself appears to invite our mind to confuse these two kinds of adaptation, for she usually begins by a passive adaptation where, later on, she will build up a mechanism for active response. Thus, in the case before us, it is unquestionable that the first rudiment of the eye is found in the pigment-spot of the lower organisms; this spot may indeed have been produced physically, by the mere action of light, and there are a great number of intermediaries between the simple spot of pigment and a complicated eye like that of the vertebrates.—But, from the fact that we pass from one thing to another by degrees, it does not follow that the two things are of the same nature. From the fact that an orator falls in, at first, with the passions of his audience in order to make himself master of them, it will not be concluded that to *follow* is the same as to *lead*. Now, living matter seems to have no other means of turning circumstances to good account than by adapting itself to them passively at the outset. Where it has to direct a movement, it begins by adopting it. Life proceeds by insinuation. The intermediate degrees between a pigment-spot and an eye are nothing to the point: however numerous the degrees, there will still be the same interval between the pigment-spot and the eye as between a photograph and a photographic apparatus. Certainly the photograph has been gradually turned into a photographic apparatus; but could light alone, a physical force, ever have provoked this change, and converted an impression left by it into a machine capable of using it?

It may be claimed that considerations of utility are out of place here; that the eye is not made to see; but that we see because we have eyes; that the organ is what it is, and "utility" is a word by which we designate the functional effects of the structure. But when I say that the eye "makes use of" light, I do not merely mean that the eye is capable of seeing; I allude to the very precise relations that exist between this organ and the apparatus of locomotion. The retina of vertebrates is prolonged in an optic nerve, which, again, is continued by cerebral centers connected with motor mechanisms. Our eye makes use of light in that it enables us to utilize, by movements of reaction, the objects that we see to be advantageous, and to avoid those which we see to be injurious. Now, of course, as light may have produced a pigment-spot by physical means, so it can physically determine the movements of certain organisms; ciliated Infusoria, for instance, react to light. But no one would hold that the influence of light has physically caused the formation of a nervous system, of a muscular system, of an osseous system, all things which are continuous with the apparatus of vision in vertebrate animals. The truth is, when one speaks of the gradual formation of the eye, and, still more, when one takes into account all that is inseparably connected with it, one brings in something entirely different from the direct action of light. One implicitly attributes to organized matter a certain capacity *sui generis,* the mysterious power of building up very complicated machines to utilize the simple excitation that it undergoes.

But this is just what is claimed to be unnecessary. Physics and chemistry are said to give us the key to everything. Eimer's great work is instructive in this respect. It is well known what persevering effort this biologist has devoted to demonstrating that transformation is brought about by the influence of the external on the internal, continuously exerted in the same direction, and not, as Darwin held, by accidental variations. His theory rests on observations of the highest interest, of which the starting-point was the study of the course followed by the color variation of the skin in certain lizards. Before this, the already old experiments of Dorfmeister had shown that the same chrysalis, according as it was submitted to cold or heat, gave rise to very different butterflies, which had long been regarded as independent species, *Vanessa levana* and *Vanessa prorsa*: an intermediate temperature produces an intermediate form. We might class with these facts the important transformations observed in a little crustacean, *Artemia salina,* when the salt of the water it lives in is increased or diminished.[1] In these various experiments the external agent seems to act as a cause of transformation. But

[1] It seems, from later observations, that the transformation of Artemia is a more complex phenomenon than was first supposed. See on this subject Samter and Heymons, "Die Variation bei Artemia Salina" (*Anhang zu den Abhandlungen der k. preussischen Akad. der Wissenschaften*, 1902).

what does the word "cause" mean here? Without undertaking an exhaustive analysis of the idea of causality, we will merely remark that three very different meanings of this term are commonly confused. A cause may act by *impelling, releasing,* or *unwinding*. The billiard ball that strikes another determines its movement by *impelling*. The spark that explodes the powder acts by *releasing*. The gradual relaxing of the spring that makes the phonograph turn *unwinds* the melody inscribed on the cylinder: if the melody which is played be the effect, and the relaxing of the spring the cause, we must say that the cause acts by *unwinding*. What distinguishes these three cases from each other is the greater or less solidarity between the cause and the effect. In the first, the quantity and quality of the effect vary with the quantity and quality of the cause. In the second, neither quality nor quantity of the effect varies with quality and quantity of the cause: the effect is invariable. In the third, the quantity of the effect depends on the quantity of the cause, but the cause does not influence the quality of the effect: the longer the cylinder turns by the action of the spring, the more of the melody I shall hear, but the nature of the melody, or of the part heard, does not depend on the action of the spring. Only in the first case, really, does cause *explain* effect; in the others the effect is more or less given in advance, and the antecedent invoked is—in different degrees, of course—its occasion rather than its cause. Now, in saying that the saltness of the water is the cause of the transformations of Artemia, or that the degree of temperature determines the color and marks of the wings which a certain chrysalis will assume on becoming a butterfly, is the word "cause" used in the first sense? Obviously not: causality has here an intermediary sense between those of unwinding and releasing. Such, indeed, seems to be Eimer's own meaning when he speaks of the "kaleidoscopic" character of the variation,[1] or when he says that the variation of organized matter works in a definite way, just as inorganic matter crystallizes in definite directions.[2] And it may be granted, perhaps, that the process is a merely physical and chemical one in the case of the color-changes of the skin. But if this sort of explanation is extended to the case of the gradual formation of the eye of the vertebrate, for instance, it must be supposed that the physico-chemistry of living bodies is such that the influence of light has caused the organism to construct a progressive series of visual apparatus, all extremely complex, yet all capable of seeing, and of seeing better and better.[3] What more could the most confirmed finalist say, in order to mark out so exceptional a physico-chemistry? And will not the position of a mechanistic philosophy become still more difficult, when it is pointed out to it that the egg of a mollusk cannot have the same chemical composition as that of a vertebrate, that the organic substance which evolved toward the first of these two forms could not have been chemically identical with that of the substance which went in the other direction, and that, nevertheless, under the influence of light, the same organ has been constructed in the one case as in the other?

The more we reflect upon it, the more we shall see that

[1] Eimer, *Orthogenesis der Schmetterlinge*, Leipzig, 1897, p. 24. Cf. *Die Entstekung der Arten*, p. 53.

[2] Eimer, *Die Entstehung der Arten*, Jena, 1888, p. 25.

[3] *Ibid.* pp. 165 ff.

this production of the same effect by two different accumulations of an enormous number of small causes is contrary to the principles of mechanistic philosophy. We have concentrated the full force of our discussion upon an example drawn from phylogenesis. But ontogenesis would have furnished us with facts no less cogent. Every moment, right before our eyes, nature arrives at identical results, in sometimes neighboring species, by entirely different embryogenic processes. Observations of "heteroblastia" have multiplied in late years,[1] and it has been necessary to reject the almost classical theory of the specificity of embryonic gills. Still keeping to our comparison between the eye of vertebrates and that of mollusks, we may point out that the retina of the vertebrate is produced by an expansion in the rudimentary brain of the young embryo. It is a regular nervous center which has moved toward the periphery. In the mollusk, on the contrary, the retina is derived from the ectoderm directly, and not indirectly by means of the embryonic encephalon. Quite different, therefore, are the evolutionary processes which lead, in man and in the Pecten, to the development of a like retina. But, without going so far as to compare two organisms so distant from each other, we might reach the same conclusion simply by looking at certain very curious facts of regeneration in one and the same organism. If the crystalline lens of a Triton be removed, it is regenerated by the iris.[2] Now, the original

[1] Salensky, "Heteroblastie" (*Proc. of the Fourth International Congress of Zoology*, London, 1899, pp. 111-118). Salensky has coined this word to designate the cases in which organs that are equivalent, but of different embryological origin, are formed at the same points in animals related to each other.

[2] Wolff, "Die Regeneration der Urodelenlinse" (*Arch. f. Entwickelungsmechanik*, i., 1895, pp. 380 ff.).

lens was built out of the ectoderm, while the iris is of mesodermic origin. What is more, in the *Salamandra maculata,* if the lens be removed and the iris left, the regeneration of the lens takes place at the upper part of the iris; but if this upper part of the iris itself be taken away, the regeneration takes place in the inner or retinal layer of the remaining region.[1] Thus, parts differently situated, differently constituted, meant normally for different functions, are capable of performing the same duties and even of manufacturing, when necessary, the same pieces of the machine. Here we have, indeed, the same effect obtained by different combinations of causes.

Whether we will or no, we must appeal to some inner directing principle in order to account for this convergence of effects. Such convergence does not appear possible in the Darwinian, and especially the neo-Darwinian, theory of insensible accidental variations, nor in the hypothesis of sudden accidental variations, nor even in the theory that assigns definite directions to the evolution of the various organs by a kind of mechanical composition of the external with the internal forces. So we come to the only one of the present forms of evolution which remains for us to mention, viz., neo-Lamarckism.

It is well known that Lamarck attributed to the living being the power of varying by use or disuse of its organs, and also of passing on the variation so acquired to its descendants. A certain number of biologists hold a doctrine of this kind today. The variation that results in a new species is not, they believe, merely an accidental

[1] Fischel, "Uber die Regeneration der Linse" (*Anat. Anzeiger*, xiv., 1898, pp. 373-380).

variation inherent in the germ itself, nor is it governed by a determinism *sui generis* which develops definite characters in a definite direction, apart from every consideration of utility. It springs from the very effort of the living being to adapt itself to the circumstances of its existence. The effort may indeed be only the mechanical exercise of certain organs, mechanically elicited by the pressure of external circumstances. But it may also imply consciousness and will, and it is in this sense that it appears to be understood by one of the most eminent representatives of the doctrine, the American naturalist Cope.[1] Neo-Lamarckism is therefore, of all the later forms of evolutionism, the only one capable of admitting an internal and psychological principle of development, although it is not bound to do so. And it is also the only evolutionism that seems to us to account for the building up of identical complex organs on independent lines of development. For it is quite conceivable that the same effort to turn the same circumstances to good account might have the same result, especially if the problem put by the circumstances is such as to admit of only one solution. But the question remains, whether the term "effort" must not then be taken in a deeper sense, a sense even more psychological than any neo-Lamarckian supposes.

For a mere variation of size is one thing, and a change of form is another. That an organ can be strengthened and grow by exercise, nobody will deny. But it is a long way from that to the progressive development of an eye like that of the mollusks and of the vertebrates. If this development be ascribed to the influence of light, long

[1] Cope, *The Origin of the Fittest*, 1887; *The Primary Factors of Organic Evolution*, 1896.

continued but passively received, we fall back on the theory we have just criticized. If, on the other hand, an internal activity is appealed to, then it must be something quite different from what we usually call an effort, for never has an effort been known to produce the slightest complication of an organ, and yet an enormous number of complications, all admirably co-ordinated, have been necessary to pass from the pigment-spot of the Infusorian to the eye of the vertebrate. But, even if we accept this notion of the evolutionary process in the case of animals, how can we apply it to plants? Here, variations of form do not seem to imply, nor always to lead to, functional changes; and even if the cause of the variation is of a psychological nature, we can hardly call it an effort, unless we give a very unusual extension to the meaning of the word. The truth is, it is necessary to dig beneath the effort itself and look for a deeper cause.

This is especially necessary, we believe, if we wish to get at a cause of regular hereditary variations. We are not going to enter here into the controversies over the transmissibility of acquired characters; still less do we wish to take too definite a side on this question, which is not within our province. But we cannot remain completely indifferent to it. Nowhere is it clearer that philosophers cannot today content themselves with vague generalities, but must follow the scientists in experimental detail and discuss the results with them. If Spencer had begun by putting to himself the question of the hereditability of acquired characters, his evolutionism would no doubt have taken an altogether different form. If (as seems probable to us) a habit contracted by the individual were transmitted to its descendants only in very exceptional cases, all the Spencerian psychology would need re-making, and a large part of Spencer's philosophy would fall to pieces. Let us say, then, how the problem seems to us to present itself, and in what direction an attempt might be made to solve it.

After having been affirmed as a dogma, the transmissibility of acquired characters has been no less dogmatically denied, for reasons drawn *a priori* from the supposed nature of germinal cells. It is well known how Weismann was led, by his hypothesis of the continuity of the germ-plasm, to regard the germinal cells—ova and spermatozoa—as almost independent of the somatic cells. Starting from this, it has been claimed, and is still claimed by many, that the hereditary transmission of an acquired character is inconceivable. But if, perchance, experiment should show that acquired characters are transmissible, it would prove thereby that the germ-plasm is not so independent of the somatic envelope as has been contended, and the transmissibility of acquired characters would become *ipso facto* conceivable; which amounts to saying that conceivability and inconceivability have nothing to do with the case, and that experience alone must settle the matter. But it is just here that the difficulty begins. The acquired characters we are speaking of are generally habits or the effects of habit, and at the root of most habits there is a natural disposition. So that one can always ask whether it is really the habit acquired by the soma of the individual that is transmitted, or whether it is not rather a natural aptitude, which existed prior to the habit. This aptitude would have remained inherent in the germ-plasm which the individual bears within him, as it was in the individual himself and consequently in the germ whence he sprang Thus, for instance, there is no proof that the mole has become blind because it has formed the habit of living underground; it is perhaps because its eyes were becoming atrophied that it condemned itself to a life underground.[1] If this is the case, the tendency to lose the power of vision has been transmitted from germ to germ without anything being acquired or lost by the soma of the mole itself. From the fact that the son of a fencing master has become a good fencer much more quickly than his father, we cannot infer that the habit of the parent has been transmitted to the child; for certain natural dispositions in course of growth may have passed from the plasma engendering the father to the plasma engendering the son, may have grown on the way by the effect of the primitive impetus, and thus assured to the son a greater suppleness than the father had, without troubling, so to speak, about what the father did. So of many examples drawn from the progressive domestication of animals: it is hard to say whether it is the acquired habit that is transmitted or only a certain natural tendency—that, indeed, which has caused such and such a particular species or certain of its representatives to be specially chosen for domestication. The truth is, when every doubtful case, every fact open to more than one interpretation, has been eliminated, there remains hardly a single unquestionable example of acquired and transmitted peculiarities, beyond the famous experiments of Brown-Séquard, repeated and confirmed by other physi-

[1] Cuénot, "La Nouvelle Théorie transformiste" (*Revue générale des sciences*, 1894). Cf. Morgan, *Evolution and Adaptation*, London, 1903, p. 357.

ologists.[1] By cutting the spinal cord or the sciatic nerve of guinea pigs, Brown-Séquard brought about an epileptic state which was transmitted to the descendants. Lesions of the same sciatic nerve, of the restiform body, etc., provoked various troubles in the guinea pig which its progeny inherited sometimes in a quite different form: exophthalmia, loss of toes, etc. But it is not demonstrated that in these different cases of hereditary transmission there had been a real influence of the soma of the animal on its germ-plasm. Weismann at once objected that the operations of Brown-Séquard might have introduced certain special microbes into the body of the guinea pig, which had found their means of nutrition in the nervous tissues and transmitted the malady by penetrating into the sexual elements.[2] This objection has been answered by Brown-Séquard himself;[3] but a more plausible one might be raised. Some experiments of Voisin and Peron have shown that fits of epilepsy are followed by the elimination of a toxic body which, when injected into animals,[4] is capable of producing convulsive symptoms. Perhaps the trophic disorders following the nerve lesions made by Brown-Séquard correspond to the formation of precisely this convulsion-causing poison. If so, the toxin passed from the guinea pig to its spermatozoon or ovum,

[1] Brown-Séquard, "Nouvelles recherches sur l'épilepsie due à certaines lésions de la moelle épiniéere et des nerfs rachidiens" (*Arch. de physiologie*, vol. ii., 1866, pp. 211, 422, and 497).

[2] Weismann, *Aufsätze über Vererbung*, Jena, 1892, pp. 376-378, and also *Vorträge über Descendenztheorie*, Jena, 1902, vol. ii., p. 76.

[3] Brown-Séquard, "Hérédité d'une affection due à une cause accidentelle" (*Arch. de physiologie*, 1892, pp. 686 ff.).

[4] Voisin and Peron, "Recherches sur la toxicité urinaire chez les épileptiques" (*Arch. de neurologie*, vol. xxiv., 1892, and xxv., 1893. Cf. the work of Voisin, *L'Épilepsie*, Paris, 1897, pp. 125-133).

and caused in the development of the embryo a general disturbance, which, however, had no visible effects except at one point or another of the organism when developed. In that case, what occurred would have been somewhat the same as in the experiments of Charrin, Delamare and Moussu, where guinea pigs in gestation, whose liver or kidney was injured, transmitted the lesion to their progeny, simply because the injury to the mother's organ had given rise to specific "cytotoxins" which acted on the corresponding organ of the foetus.[1] It is true that, in these experiments, as in a former observation of the same physiologists,[2] it was the already formed foetus that was influenced by the toxins. But other researches of Charrin have resulted in showing that the same effect may be produced, by an analogous process, on the spermatozoa and the ova.[3] To conclude, then: the inheritance of an acquired peculiarity in the experiments of Brown-Séquard can be explained by the effect of a toxin on the germ. The lesion, however well localized it seems, is transmitted by the same process as, for instance, the taint of alcoholism. But may it not be the same in the case of every acquired peculiarity that has become hereditary?

There is, indeed, one point on which both those who affirm and those who deny the transmissibility of ac-

[1] Charrin, Delamare and Moussu, "Transmission expérimentale aux descendants de lésions développées chez les ascendants" (*C. R. de l'Acad. des sciences*, vol. cxxxv., 1902, p. 191). Cf. Morgan, *Evolution and Adaptation*, p. 257, and Delage, *L'Hérédité*, 2nd edition, p. 388.

[2] Charrin ad Delamare, "Hérédité cellulaire" (*C. R. de l'Acad. des sciences*, vol. cxxxiii., 1901, pp. 69-71).

[3] Charrin, "L'Hérédité pathologique" (*Revue générale des sciences*, 15 janvier 1896).

quired characters are agreed, namely, that certain influences, such as that of alcohol, can affect at the same time both the living being and the germ-plasm it contains. In such case, there is inheritance of a defect, and the result is *as if* the soma of the parent had acted on the germ-plasm, although in reality soma and plasma have simply both suffered the action of the same cause. Now, suppose that the soma can influence the germ-plasm, as those believe who hold that acquired characters are transmissible. Is not the most natural hypothesis to suppose that things happen in this second case as in the first, and that the direct effect of the influence of the soma is a *general* alteration of the germ-plasm? If this is the case, it is by exception, and in some sort by accident, that the modification of the descendant is the same as that of the parent. It is like the hereditability of the alcoholic taint: it passes from father to children, but it may take a different form in each child, and in none of them be like what it was in the father. Let the letter C represent the change in the plasm, C being either positive or negative, that is to say, showing either the gain or loss of certain substances. The effect will not be an exact reproduction of the cause, nor will the change in the germ-plasm, provoked by a certain modification of a certain part of the soma, determine a similar modification of the corresponding part of the new organism in process of formation, unless all the other nascent parts of this organism enjoy a kind of immunity as regards C: the same part will then undergo alteration in the new organism, because it happens that the development of this part is alone subject to the new influence. And, even then, the part might be altered in an entirely different way from that in which the corresponding part was altered in the generating organism.

We should propose, then, to introduce a distinction between the hereditability of *deviation* and that of *character*. An individual which acquires a new character thereby *deviates* from the form it previously had, which form the germs, or oftener the half-germs, it contains would have reproduced in their development. If this modification does not involve the production of substances capable of changing the germ-plasm, or does not so affect nutrition as to deprive the germ-plasm of certain of its elements, it will have no effect on the offspring of the individual. This is probably the case as a rule. If, on the contrary, it has some effect, this is likely to be due to a chemical change which it has induced in the germ-plasm. This chemical change might, by exception, bring about the original modification again in the organism which the germ is about to develop, but there are as many and more chances that it will do something else. In this latter case, the generated organism will perhaps deviate from the normal type *as much as* the generating organism, but it will do so *differently*. It will have inherited deviation and not character. In general, therefore, the habits formed by an individual have probably no echo in its offspring; and when they have, the modification in the descendants may have no visible likeness to the original one. Such, at least, is the hypothesis which seems to us most likely. In any case, in default of proof to the contrary, and so long as the decisive experiments called for by an eminent biologist[1] have not been made, we must keep to the actual results of ob-

[1] Giard, *Controverses transformistes* Paris, 1904, p. 147.

servation. Now, even if we take the most favorable view of the theory of the transmissibility of acquired characters, and assume that the ostensible acquired character is not, in most cases, the more or less tardy development of an innate character, facts show us that hereditary transmission is the exception and not the rule. How, then, shall we expect it to develop an organ such as the eye? When we think of the enormous number of variations, all in the same direction, that we must suppose to be accumulated before the passage from the pigment-spot of the Infusorian to the eye of the mollusk and of the vertebrate is possible, we do not see how heredity, as we observe it, could ever have determined this piling-up of differences, even supposing that individual efforts could have produced each of them singly. That is to say that neo-Lamarckism is no more able than any other form of evolutionism to solve the problem.

In thus submitting the various present forms of evolutionism to a common test, in showing that they all strike against the same insurmountable difficulty, we have in no wise the intention of rejecting them altogether. On the contrary, each of them, being supported by a considerable number of facts, must be true in its way. Each of them must correspond to a certain aspect of the process of evolution. Perhaps even it is necessary that a theory should restrict itself exclusively to a particular point of view, in order to remain scientific, *i.e.* to give a precise direction to researches into detail. But the reality of which each of these theories takes a partial view must transcend them all. And this reality is the special object of philosophy, which is not constrained to scientific precision because it contemplates no practical application. Let us therefore indicate in a word or two the positive contribution that each of the three present forms of evolutionism seems to us to make toward the solution of the problem, what each of them leaves out, and on what point this threefold effort should, in our opinion, converge in order to obtain a more comprehensive, although thereby of necessity a less definite, idea of the evolutionary process.

The neo-Darwinians are probably right, we believe, when they teach that the essential causes of variation are the differences inherent in the germ borne by the individual, and not the experiences or behavior of the individual in the course of his career. Where we fail to follow these biologists, is in regarding the differences inherent in the germ as purely accidental and individual. We cannot help believing that these differences are the development of an impulsion which passes from germ to germ across the individuals, that they are therefore not pure accidents, and that they might well appear at the same time, in the same form, in all the representatives of the same species, or at least in a certain number of them. Already, in fact, the theory of *mutations* is modifying Darwinism profoundly on this point. It asserts that at a given moment, after a long period, the entire species is beset with a tendency to change. The *tendency to change*, therefore, is not accidental. True, the change itself would be accidental, since the mutation works, according to De Vries, in different directions in the different representatives of the species. But, first we must see if the theory is confirmed by many other vegetable species (De Vries has verified it only by the *Œnothera Lamarck-*

iana),[1] and then there is the possibility, as we shall explain further on, that the part played by chance is much greater in the variation of plants than in that of animals, because, in the vegetable world, function does not depend so strictly on form. Be that as it may, the neo-Darwinians are inclined to admit that the periods of mutation are determinate. The direction of the mutation may therefore be so as well, at least in animals, and to the extent we shall have to indicate.

We thus arrive at a hypothesis like Eimer's, according to which the variations of different characters continue from generation to generation in definite directions. This hypothesis seems plausible to us, within the limits in which Eimer himself retains it. Of course, the evolution of the organic world cannot be predetermined as a whole. We claim, on the contrary, that the spontaneity of life is manifested by a continual creation of new forms succeeding others. But this indetermination cannot be complete; it must leave a certain part to determination. An organ like the eye, for example, must have been formed by just a continual changing in a definite direction. Indeed, we do not see how otherwise to explain the likeness of structure of the eye in species that have not the same history. Where we differ from Eimer is in his claim that combinations of physical and chemical causes are enough to secure the result. We have tried to prove, on the contrary, by the example of the eye,

[1] Some analogous facts, however, have been noted, all in the vegetable world. See Blaringhem, "La Notion d'espèce et la théorie de la mutation" (*Année psychologique*, vol. xii., 1906, pp. 95 ff.), and De Vries, *Species and Varieties*, p. 655.

that if there is "orthogenesis" here, a psychological cause intervenes.

Certain neo-Lamarckians do indeed resort to a cause of a psychological nature. There, to our thinking, is one of the most solid positions of neo-Lamarckism. But if this cause is nothing but the conscious effort of the individual, it cannot operate in more than a restricted number of cases—at most in the animal world, and not at all in the vegetable kingdom. Even in animals, it will act only on points which are under the direct or indirect control of the will. And even where it does act, it is not clear how it could compass a change so profound as an increase of complexity: at most this would be conceivable if the acquired characters were regularly transmitted so as to be added together; but this transmission seems to be the exception rather than the rule. A hereditary change in a definite direction, which continues to accumulate and add to itself so as to build up a more and more complex machine, must certainly be related to some sort of effort, but to an effort of far greater depth than the individual effort, far more independent of circumstances, an effort common to most representatives of the same species, inherent in the germs they bear rather than in their substance alone, an effort thereby assured of being passed on to their descendants.

So we come back, by a somewhat roundabout way, to the idea we started from, that of an *original impetus* of life, passing from one generation of germs to the following generation of germs through the developed organisms which bridge the interval between the generations.

This impetus, sustained right along the lines of evolution among which it gets divided, is the fundamental cause of variations, at least of those that are regularly passed on, that accumulate and create new species. In general, when species have begun to diverge from a common stock, they accentuate their divergence as they progress in their evolution. Yet, in certain definite points, they may evolve identically; in fact, they must do so if the hypothesis of a common impetus be accepted. This is just what we shall have to show now in a more precise way, by the same example we have chosen, the formation of the eye in mollusks and vertebrates. The idea of an "original impetus," moreover, will thus be made clearer.

Two points are equally striking in an organ like the eye: the complexity of its structure and the simplicity of its function. The eye is composed of distinct parts, such as the sclerotic, the cornea, the retina, the crystalline lens, etc. In each of these parts the detail is infinite. The retina alone comprises three layers of nervous elements—multipolar cells, bipolar cells, visual cells—each of which has its individuality and is undoubtedly a very complicated organism: so complicated, indeed, is the retinal membrane in its intimate structure, that no simple description can give an adequate idea of it. The mechanism of the eye is, in short, composed of an infinity of mechanisms, all of extreme complexity. Yet vision is one simple fact. As soon as the eye opens, the visual act is effected. Just because the act is simple, the slightest negligence on the part of nature in the building of the infinitely complex machine would have made vision impossible. This contrast between the complexity of the organ and the unity of the function is what gives us pause.

A mechanistic theory is one which means to show us the gradual building-up of the machine under the influence of external circumstances intervening either directly by action on the tissues or indirectly by the selection of better-adapted ones. But, whatever form this theory may take, supposing it avails at all to explain the detail of the parts, it throws no light on their correlation.

Then comes the doctrine of finality, which says that the parts have been brought together on a preconceived plan with a view to a certain end. In this it likens the labor of nature to that of the workman, who also proceeds by the assemblage of parts with a view to the realization of an idea or the imitation of a model. Mechanism, here, reproaches finalism with its anthropomorphic character, and rightly. But it fails to see that itself proceeds according to this method—somewhat mutilated! True, it has got rid of the end pursued or the ideal model. But it also holds that nature has worked like a human being by bringing parts together, while a mere glance at the development of an embryo shows that life goes to work in a very different way. *Life does not proceed by the association and addition of elements, but by dissociation and division.*

We must get beyond both points of view, both mechanism and finalism being, at bottom, only standpoints to which the human mind has been led by considering the work of man. But in what direction can we go beyond them? We have said that in analyzing the structure of an organ, we can go on decomposing forever, although the function of the whole is a simple thing. This contrast between the infinite complexity of the organ and the extreme simplicity of the function is what should open our eyes.

In general, when the same object appears in one aspect as simple and in another as infinitely complex, the two aspects have by no means the same importance, or rather the same degree of reality. In such cases, the simplicity belongs to the object itself, and the infinite complexity to the views we take in turning around it, to the symbols by which our senses or intellect represent it to us, or, more generally, to elements *of a different order*, with which we try to imitate it artificially, but with which it remains incommensurable, being of a different nature. An artist of genius has painted a figure on his canvas. We can imitate his picture with many-colored squares of mosaic. And we shall reproduce the curves and shades of the model so much the better as our squares are smaller, more numerous and more varied in tone. But an infinity of elements infinitely small, presenting an infinity of shades, would be necessary to obtain the exact equivalent of the figure that the artist has conceived as a simple thing, which he has wished to transport as a whole to the canvas, and which is the more complete the more it strikes us as the projection of an indivisible intuition. Now, suppose our eyes so made that they cannot help seeing in the work of the master a mosaic effect. Or suppose our intellect so made that it cannot explain the appearance of the figure on the canvas except as a work of mosaic. We should then be able to speak simply of a collection of little squares, and we should be under the mechanistic hypothesis. We might add that, besides the materiality of the collection, there must be a plan on which the artist worked; and then we should be expressing ourselves as finalists. But in neither case should we have got at the real process, for there are no squares brought together. It is the picture, *i.e.* the simple act, projected on the canvas, which, by the mere fact of entering into our perception, is *de*composed before our eyes into thousands and thousands of little squares which present, as *re*composed, a wonderful arrangement. So the eye, with its marvelous complexity of structure, may be only the simple act of vision, divided *for us* into a mosaic of cells, whose order seems marvelous to us because we have conceived the whole as an assemblage.

If I raise my hand from A to B, this movement appears to me under two aspects at once. Felt from within, it is a simple, indivisible act. Perceived from without, it is the course of a certain curve, AB. In this curve I can distinguish as many positions as I please, and the line itself might be defined as a certain mutual co-ordination of these positions. But the positions, infinite in number, and the order in which they are connected, have sprung automatically from the indivisible act by which my hand has gone from A to B. Mechanism, here, would consist in seeing only the positions. Finalism would take their order into account. But both mechanism and finalism would leave on one side the movement, which is reality itself. In one sense, the movement is *more* than the positions and than their order; for it is sufficient to make it in its indivisible simplicity to secure that the infinity of the successive positions as also their order be given at once—with something else which is neither order nor position but which is essential, the mobility. But, in another sense, the movement is *less* than the series of positions and their connecting order; for, to arrange points in a certain order, it is necessary first to conceive the order and then to realize it with points; there must be the work of assemblage and there must be intelligence, whereas the simple movement of the hand contains nothing of either. It is not intelligent, in the human sense of the word, and it is not an assemblage, for it is not made up of elements. Just so with the relation of the eye to vision. There is in vision *more* than the component cells of the eye and their mutual co-ordination: in this sense, neither mechanism nor finalism go far enough. But, in another sense, mechanism and finalism both go too far, for they attribute to Nature the most formidable of the labors of Hercules in holding that she has exalted to the simple act of vision an infinity of infinitely complex elements, whereas Nature has had no more trouble in making an eye than I have in lifting my hand. Nature's simple act has divided itself automatically into an infinity of elements which are then found to be co-ordinated to one idea, just as the movement of my hand has dropped an infinity of points which are then found to satisfy one equation.

We find it very hard to see things in that light, because we cannot help conceiving organization as manufacturing. But it is one thing to manufacture, and quite another to organize. Manufacturing is peculiar to man. It consists in assembling parts of matter which we have cut out in such manner that we can fit them together and obtain from them a common action. The parts are arranged, so to speak, around the action as an ideal center. To manufacture, therefore, is to work from the periphery to the center, or, as the philosophers say, from the many to the one. Organization, on the contrary, works from the center to the periphery. It begins in a point that is almost a mathematical point, and spreads around this point by concentric waves which go on enlarging. The work of manufacturing is the more effective, the greater the quantity of matter dealt with. It proceeds by concentration and compression. The organizing act, on the contrary, has something explosive about it: it needs at the beginning the smallest possible place, a minimum of matter, as if the organizing forces only entered space reluctantly. The spermatozoon, which sets in motion the evolutionary process of the embryonic life, is one of the smallest cells of the organism; and it is only a small part of the spermatozoon which really takes part in the operation.

But these are only superficial differences. Digging beneath them, we think, a deeper difference would be found.

A manufactured thing delineates exactly the form of the work of manufacturing it. I mean that the manufacturer finds in his product exactly what he has put into it. If he is going to make a machine, he cuts out its pieces one by one and then puts them together: the machine, when made, will show both the pieces and their assemblage. The whole of the result represents the whole of the work; and to each part of the work corresponds a part of the result.

Now I recognize that positive science can and should proceed as if organization was like making a machine. Only so will it have any hold on organized bodies. For its object is not to show us the essence of things, but to furnish us with the best means of acting on them. Physics and chemistry are well advanced sciences, and living matter lends itself to our action only so far as we can treat it by the processes of our physics and chemistry. Organization can therefore only be studied scientifically if the organized body has first been likened to a machine. The cells will be the pieces of the machine, the organism their assemblage, and the elementary labors which have organized the parts will be regarded as the real elements of the labor which has organized the whole. This is the standpoint of science. Quite different, in our opinion, is that of philosophy.

For us, the whole of an organized machine may, strictly speaking, represent the whole of the organizing work (this is, however, only approximately true), yet the parts of the machine do not correspond to parts of the work, because *the materiality of this machine does not represent a sum of means employed, but a sum of obstacles avoided*: it is a negation rather than a positive reality. So, as we have shown in a former study, vision is a power which should attain *by right* an infinity of things inaccessible to our eyes. But such a vision would not be continued into action; it might suit a phantom, but not a living being. The vision of a living being is an *effective* vision, limited to objects on which the being can act: it is a vision that is *canalized*, and the visual apparatus simply symbolizes the work of canalizing. Therefore the creation of the visual apparatus is no more explained by the assembling of its anatomic elements than the digging of a canal could be explained by the heaping-up of the earth which might have formed its banks. A mechanistic theory would maintain that the earth had been brought cart-load by cart-load; finalism would add that it had not been dumped down at random, that the carters had followed a plan. But both theories would be mistaken, for the canal has been made in another way.

With greater precision, we may compare the process by which nature constructs an eye to the simple act by which we raise the hand. But we supposed at first that the hand met with no resistance. Let us now imagine that, instead of moving in air, the hand has to pass through iron filings which are compressed and offer resistance to it in proportion as it goes forward. At a certain moment the hand will have exhausted its effort, and, at this very moment, the filings will be massed and coordinated in a certain definite form, to wit, that of the hand that is stopped and of a part of the arm. Now, suppose that the hand and arm are invisible. Lookers-on will seek the reason of the arrangement in the filings themselves and in forces within the mass. Some will account for the position of each filing by the action exerted upon it by the neighboring filings: these are the mechanists. Others will prefer to think that a plan of the whole has presided over the detail of these elementary actions: they are the finalists. But the truth is that there has been merely one indivisible act, that of the hand passing through the filings: the inexhaustible detail of the movement of the grains, as well as the order of their final arrangement, expresses negatively, in a way, this undivided movement, being the unitary form of a resistance, and not a synthesis of positive elementary actions. For this reason, if the arrangement of the grains is termed an "effect" and the movement of the hand a "cause," it may indeed be said that the whole of the effect is explained by the whole of the cause, but to parts of the cause parts of the effect will in no wise correspond. In other words, neither mechanism nor finalism will here be in place, and we must resort to an explanation of a different kind. Now, in the hypothesis we propose, the relation of vision to the visual apparatus would be very nearly that of the hand to the iron filings that follow, canalize and limit its motion.

The greater the effort of the hand, the farther it will go into the filings. But at whatever point it stops, instantaneously and automatically the filings co-ordinate and find their equilibrium. So with vision and its organ. According as the undivided act constituting vision advances more or less, the materiality of the organ is made of a more or less considerable number of mutually co-ordinated elements, but the order is necessarily complete and perfect. It could not be partial, because, once again, the real process which gives rise to it has no parts. That is what neither mechanism nor finalism takes into account, and it is what we also fail to consider when we wonder at the marvelous structure of an instrument such as the eye. At the bottom of our wondering is always this idea, that it would have been possible for *a part only* of this co-ordination to have been realized, that the complete realization is a kind of special favor. This favor the finalists consider as dispensed to them all at once, by the final cause; the mechanists claim to obtain it little by little, by the effect of natural selection; but both see something positive in this co-ordination, and consequently something fractionable in its cause—something which admits of every possible degree of achievement. In reality, the cause, though more or less intense, cannot produce its effect except in one piece, and completely finished. According as it goes further and further in the direction of vision, it gives the simple pigmentary masses of a lower organism, or the rudimentary eye of a Serpula, or the slightly differentiated eye of the Alciope, or the marvelously perfected eye of the bird; but all these organs, unequal as is their complexity, necessarily present an equal co-ordination. For this reason, no matter how distant two animal species may be from each other, if the progress toward vision has gone equally far in both, there is the same visual organ in each case, for the form of the organ only expresses the degree in which the exercise of the function has been obtained.

But, in speaking of a progress toward vision, are we not coming back to the old notion of finality? It would be so, undoubtedly, if this progress required the conscious or unconscious idea of an end to be attained. But it is really effected in virtue of the original impetus of life; it is implied in this movement itself, and that is just why it is found in independent lines of evolution. If now we are asked why and how it is implied therein, we reply that life is, more than anything else, a tendency to act on inert matter. The direction of this action is not predetermined; hence the unforeseeable variety of forms which life, in evolving, sows along its path. But this action always presents, to some extent, the character of contingency; it implies at least a rudiment of choice. Now a choice involves the anticipatory idea of several possible actions. Possibilities of action must therefore be marked out for the living being before the action itself. Visual perception is nothing else:[1] the visible outlines of

[1] See, on this subject, *Matière et mémoire*, chap. i.

bodies are the design of our eventual action on them. Vision will be found, therefore, in different degrees in the most diverse animals, and it will appear in the same complexity of structure wherever it has reached the same degree of intensity.

We have dwelt on these resemblances of structure in general, and on the example of the eye in particular, because we had to define our attitude toward mechanism on the one hand and finalism on the other. It remains for us to describe it more precisely in itself. This we shall now do by showing the divergent results of evolution not as presenting analogies, but as themselves mutually complementary.

CHAPTER II

THE DIVERGENT DIRECTIONS OF THE EVOLUTION OF LIFE. TORPOR, INTELLIGENCE, INSTINCT

THE evolution movement would be a simple one, and we should soon have been able to determine its direction, if life had described a single course, like that of a solid ball shot from a cannon. But it proceeds rather like a shell, which suddenly bursts into fragments, which fragments, being themselves shells, burst in their turn into fragments destined to burst again, and so on for a time incommensurably long. We perceive only what is nearest to us, namely, the scattered movements of the pulverized explosions. From them we have to go back, stage by stage, to the original movement.

When a shell bursts, the particular way it breaks is explained both by the explosive force of the powder it contains and by the resistance of the metal. So of the way life breaks into individuals and species. It depends, we think, on two series of causes: the resistance life meets from inert matter, and the explosive force—due to an unstable balance of tendencies—which life bears within itself.

The resistance of inert matter was the obstacle that had first to be overcome. Life seems to have succeeded in this by dint of humility, by making itself very small and very insinuating, bending to physical and chemical forces, consenting even to go a part of the way with them, like the switch that adopts for a while the direction of the rail it is endeavoring to leave. Of phenomena in the simplest forms of life, it is hard to say whether they are still physical and chemical or whether they are already vital. Life had to enter thus into the habits of inert matter, in order to draw it little by little, magnetized, as it were, to another track. The animate forms that first appeared were therefore of extreme simplicity. They were probably tiny masses of scarcely differentiated protoplasm, outwardly resembling the amoeba observable today, but possessed of the tremendous internal push that was to raise them even to the highest forms of life. That in virtue of this push the first organisms sought to grow as much as possible, seems likely. But organized matter has a limit of expansion that is very quickly reached; beyond a certain point it divides instead of growing. Ages of effort and prodigies of subtlety were probably necessary for life to get past this new obstacle. It succeeded in inducing an increasing number of elements, ready to divide, to remain united. By the division of labor it knotted between them an indissoluble bond. The complex and quasi-discontinuous organism is thus made to function as would a continuous living mass which had simply grown bigger.

But the real and profound causes of division were those which life bore within its bosom. For life is tendency, and the essence of a tendency is to develop in the form of a sheaf, creating, by its very growth, divergent directions among which its impetus is divided. This we observe in ourselves, in the evolution of that special tendency which we call our character. Each of us, glancing back over his history, will find that his child-personality, though indivisible, united in itself divers persons, which could remain blended just because they were in their nascent state: this indecision, so charged with promise, is one of the greatest charms of childhood. But these interwoven personalities become incompatible in course of growth, and, as each of us can live but one life, a choice must perforce be made. We choose in reality without ceasing; without ceasing, also, we abandon many things. The route we pursue in time is strewn with the remains of all that we began to be, of all that we might have become. But nature, which has at command an incalculable number of lives, is in no wise bound to make such sacrifices. She preserves the different tendencies that have bifurcated with their growth. She creates with them diverging series of species that will evolve separately.

These series may, moreover, be of unequal importance. The author who begins a novel puts into his hero many things which he is obliged to discard as he goes on. Perhaps he will take them up later in other books, and make new characters with them, who will seem like extracts from, or rather like complements of, the first; but they will almost always appear somewhat poor and limited in comparison with the original character. So with regard to the evolution of life. The bifurcations on the way have been numerous, but there have been many blind alleys beside the two or three highways; and of these highways themselves, only one, that which leads through the vertebrates up to man, has been wide enough to allow free passage to the full breath of life. We get this impression when we compare the societies of bees and ants, for instance, with human societies. The former are admirably ordered and united, but stereotyped; the latter are open to every sort of progress, but divided, and incessantly at strife with themselves. The ideal would be a society always in progress and always in equilibrium, but this ideal is perhaps unrealizable: the two characteristics that would fain complete each other, which do complete each other in their embryonic state, can no longer abide together when they grow stronger. If one could speak, otherwise than metaphorically, of an impulse toward social life, it might be said that the brunt of the impulse was borne along the line of evolution ending at man, and that the rest of it was collected on the road leading to the hymenoptera: the societies of ants and bees would thus present the aspect complementary to ours. But this would be only a manner of expression. There has been no particular impulse toward social life; there is simply the general movement of life, which on divergent lines is creating forms ever new. If societies should appear on two of these lines, they ought to show divergence of paths at the same time as community of impetus. They will thus develop two classes of characteristics which we shall find vaguely complementary of each other.

So our study of the evolution movement will have to unravel a certain number of divergent directions, and to appreciate the importance of what has happened along each of them—in a word, to determine the nature of the dissociated tendencies and estimate their relative proportion. Combining these tendencies, then, we shall get an approximation, or rather an imitation, of the indivisible motor principle whence their impetus proceeds. Evolution will thus prove to be something entirely different from a series of adaptations to circumstances, as mechanism claims; entirely different also from the realization of a plan of the whole, as maintained by the doctrine of finality.

That adaptation to environment is the necessary condition of evolution we do not question for a moment. It is quite evident that a species would disappear, should it fail to bend to the conditions of existence which are imposed on it. But it is one thing to recognize that outer circumstances are forces evolution must reckon with, another to claim that they are the directing causes of evolution. This latter theory is that of mechanism. It excludes absolutely the hypothesis of an original impetus, I mean an internal push that has carried life, by more and more complex forms, to higher and higher destinies. Yet this impetus is evident, and a mere glance at fossil species shows us that life need not have evolved at all, or might have evolved only in very restricted limits, if it had chosen the alternative, much more convenient to itself, of becoming anchylosed in its primitive forms. Certain Foraminifera have not varied since the Silurian epoch. Unmoved witnesses of the innumerable revolutions that have upheaved our planet, the Lingulae are today what they were at the remotest times of the paleozic era.

The truth is that adaptation explains the sinuosities of the movement of evolution, but not its general directions, still less the movement itself.[1] The road that leads to the town is obliged to follow the ups and downs of the hills; it *adapts itself* to the accidents of the ground; but the accidents of the ground are not the cause of the road, nor have they given it its direction. At every moment they furnish it with what is indispensable, namely, the soil on which it lies; but if we consider the whole of the road, instead of each of its parts, the accidents of the ground appear only as impediments or causes of delay, for the road aims simply at the town and would fain be a straight line. Just so as regards the evolution of life and the circumstances through which it passes—with this difference, that evolution does not mark out a solitary route, that it takes directions without aiming at ends, and that it remains inventive even in its adaptations.

But, if the evolution of life is something other than a series of adaptations to accidental circumstances, so also it is not the realization of a plan. A plan is given in advance. It is represented, or at least representable, before its realization. The complete execution of it may be put off to a distant future, or even indefinitely; but the idea is none the less formulable at the present time, in terms actually given. If, on the contrary, evolution is a creation unceasingly renewed, it creates, as it goes on, not only the forms of life, but the ideas that will enable the intellect to understand it, the terms which will serve to express it. That is to say that its future overflows its present, and cannot be sketched out therein in an idea.

[1] This view of adaptation has been noted by M. F. Marin in a remarkable article on the origin of species, "L'Origine des espèces" (*Revue scientifique*, Nov. 1901, p. 580).

There is the first error of finalism. It involves another, yet more serious.

If life realizes a plan, it ought to manifest a greater harmony the further it advances, just as the house shows better and better the idea of the architect as stone is set upon stone. If, on the contrary, the unity of life is to be found solely in the impetus that pushes it along the road of time, the harmony is not in front, but behind. The unity is derived from a *vis a tergo*: it is given at the start as an impulsion, not placed at the end as an attraction. In communicating itself, the impetus splits up more and more. Life, in proportion to its progress, is scattered in manifestations which undoubtedly owe to their common origin the fact that they are complementary to each other in certain aspects, but which are none the less mutually incompatible and antagonistic. So the discord between species will go on increasing. Indeed, we have as yet only indicated the essential cause of it. We have supposed, for the sake of simplicity, that each species received the impulsion in order to pass it on to others, and that, in every direction in which life evolves, the propagation is in a straight line. But, as a matter of fact, there are species which are arrested; there are some that retrogress. Evolution is not only a movement forward; in many cases we observe a marking-time, and still more often a deviation or turning back. It must be so, as we shall show further on, and the same causes that divide the evolution movement often cause life to be diverted from itself, hypnotized by the form it has just brought forth. Thence results an increasing disorder. No doubt there is progress, if progress mean a continual advance in the general direction determined by a first impulsion; but this progress is accomplished only on the two or three great lines of evolution on which forms ever more and more complex, ever more and more high, appear; between these lines run a crowd of minor paths in which, on the contrary, deviations, arrests, and set-backs, are multiplied. The philosopher, who begins by laying down as a principle that each detail is connected with some general plan of the whole, goes from one disappointment to another as soon as he comes to examine the facts; and, as he had put everything in the same rank, he finds that, as the result of not allowing for accident, he must regard everything as accidental. For accident, then, an allowance must first be made, and a very liberal allowance. We must recognize that all is not coherent in nature. By so doing, we shall be led to ascertain the centers around which the incoherence crystallizes. This crystallization itself will clarify the rest; the main directions will appear, in which life is moving whilst developing the original impulse. True, we shall not witness the detailed accomplishment of a plan. Nature is more and better than a plan in course of realization. A plan is a term assigned to a labor: it closes the future whose form it indicates. Before the evolution of life, on the contrary, the portals of the future remain wide open. It is a creation that goes on forever in virtue of an initial movement. This movement constitutes the unity of the organized world—a prolific unity, of an infinite richness, superior to any that the intellect could dream of, for the intellect is only one of its aspects or products.

But it is easier to define the method than to apply it. The complete interpretation of the evolution movement in the past, as we conceive it, would be possible only if the history of the development of the organized world were entirely known. Such is far from being the case. The genealogies proposed for the different species are generally questionable. They vary with their authors, with the theoretic views inspiring them, and raise discussions to which the present state of science does not admit of a final settlement. But a comparison of the different solutions shows that the controversy bears less on the main lines of the movement than on matters of detail; and so, by following the main lines as closely as possible, we shall be sure of not going astray. Moreover, they alone are important to us; for we do not aim, like the naturalist, at finding the order of succession of different species, but only at defining the principal directions of their evolution. And not all of these directions have the same interest for us: what concerns us particularly is the path that leads to man. We shall therefore not lose sight of the fact, in following one direction and another, that our main business is to determine the relation of man to the animal kingdom, and the place of the animal kingdom itself in the organized world as a whole.

To begin with the second point, let us say that no definite characteristic distinguishes the plant from the animal. Attempts to define the two kingdoms strictly have always come to naught. There is not a single property of vegetable life that is not found, in some degree, in certain animals; not a single characteristic feature of the animal that has not been seen in certain species or at certain moments in the vegetable world. Naturally, therefore, biologists enamored of clean-cut concepts have regarded the distinction between the two kingdoms as artificial. They would be right, if definition in this case must be made, as in the mathematical and physical sciences, according to certain statical attributes which belong to the object defined and are not found in any other. Very different, in our opinion, is the kind of definition which befits the sciences of life. There is no manifestation of life which does not contain, in a rudimentary state—either latent or potential,—the essential characters of most other manifestations. The difference is in the proportions. But this very difference of proportion will suffice to define the group, if we can establish that it is not accidental, and that the group, as it evolves, tends more and more to emphasize these particular characters. In a word, *the group must not be defined by the possession of certain characters, but by its tendency to emphasize them.* From this point of view, taking tendencies rather than states into account, we find that vegetables and animals may be precisely defined and distinguished, and that they correspond to two divergent developments of life.

This divergence is shown, first, in the method of alimentation. We know that the vegetable derives directly from the air and water and soil the elements necessary to maintain life, especially carbon and nitrogen, which it takes in mineral form. The animal, on the contrary, cannot assimilate these elements unless they have already been fixed for it in organic substances by plants, or by animals which directly or indirectly owe them to plants; so that ultimately the vegetable nourishes the animal. True, this law allows of many exceptions among vegetables. We do not hesitate to class amongst vegetables the Drosera, the Dionaea, the Pinguicula, which are insectivorous plants. On the other hand, the fungi, which occupy so considerable a place in the vegetable world, feed like animals: whether they are ferments, saprophytes or parasites, it is to already formed organic substances that they owe their nourishment. It is therefore impossible to draw from this difference any *static* definition such as would automatically settle in any particular case the question whether we are dealing with a plant or an animal. But the difference may provide the beginning of a *dynamic* definition of the two kingdoms, in that it marks the two divergent directions in which vegetables and animals have taken their course. It is a remarkable fact that the fungi, which nature has spread all over the earth in such extraordinary profusion, have not been able to evolve. Organically they do not rise above tissues which, in the higher vegetables, are formed in the embryonic sac of the ovary, and precede the germinative development of the new individual.[1] They might be called the abortive children of the vegetable world. Their different species are like so many blind alleys, as if, by renouncing the mode of alimentation customary amongst vegetables, they had been brought to a standstill on the highway of vegetable evolution. As to the Drosera, the Dionaea, and insectivorous plants in general, they are fed by their roots, like other plants; they too fix, by their green parts, the carbon of the carbonic acid in the atmosphere. Their faculty of capturing, absorbing and digesting insects must have arisen late, in quite exceptional cases where the soil was too poor to furnish sufficient nourishment. In a general way, then,

[1] De Saporta and Marion, *L'Évolution des cryptogames*, 1881, p. 37.

if we attach less importance to the presence of special characters than to their tendency to develop, and if we regard as essential that tendency along which evolution has been able to continue indefinitely, we may say that vegetables are distinguished from animals by their power of creating organic matter out of mineral elements which they draw directly from the air and earth and water. But now we come to another difference, deeper than this, though not unconnected with it.

The animal, being unable to fix directly the carbon and nitrogen which are everywhere to be found, has to seek for its nourishment vegetables which have already fixed these elements, or animals which have taken them from the vegetable kingdom. So the animal must be able to move. From the amoeba, which thrusts out its pseudopodia at random to seize the organic matter scattered in a drop of water, up to the higher animals which have sense-organs with which to recognize their prey, locomotor organs to go and seize it, and a nervous system to co-ordinate their movements with their sensations, animal life is characterized, in its general direction, by mobility in space. In its most rudimentary form, the animal is a tiny mass of protoplasm enveloped at most in a thin albuminous pellicle which allows full freedom for change of shape and movement. The vegetable cell, on the contrary, is surrounded by a membrane of cellulose, which condemns it to immobility. And, from the bottom to the top of the vegetable kingdom, there are the same habits growing more and more sedentary, the plant having no need to move, and finding around it, in the air and water and soil in which it is placed, the mineral elements it can appropriate directly. It is true that phenomena of movement are seen in plants. Darwin has written a well-known work on the movements of climbing plants. He studied also the contrivances of certain insectivorous plants, such as the Drosera and the Dionaea, to seize their prey. The leaf-movements of the acacia, the sensitive plant, etc., are well known. Moreover, the circulation of the vegetable protoplasm within its sheath bears witness to its relationship to the protoplasm of animals, whilst in a large number of animal species (generally parasites) phenomena of fixation, analogous to those of vegetables, can be observed.[1] Here, again, it would be a mistake to claim that fixity and mobility are the two characters which enable us to decide, by simple inspection alone, whether we have before us a plant or an animal. But fixity, in the animal, generally seems like a torpor into which the species has fallen, a refusal to evolve further in a certain direction; it is closely akin to parasitism and is accompanied by features that recall those of vegetable life. On the other hand, the movements of vegetables have neither the frequency nor the variety of those of animals. Generally, they involve only part of the organism and scarcely ever extend to the whole. In the exceptional cases in which a vague spontaneity appears in vegetables, it is as if we beheld the accidental awakening of an activity normally asleep. In short, although both mobility and fixity exist in the vegetable as in the animal world, the balance is clearly in favor of fixity in the one case and of mobility in the other. These two opposite tendencies are so plainly directive of the two evolutions that the two kingdoms

[1] On fixation and parasitism in general, see the work of Houssay, *La Forme et la vie*, Paris, 1900, pp. 721-807.

might almost be defined by them. But fixity and mobility, again, are only superficial signs of tendencies that are still deeper.

Between mobility and consciousness there is an obvious relationship. No doubt, the consciousness of the higher organisms seems bound up with certain cerebral arrangements. The more the nervous system develops, the more numerous and more precise become the movements among which it can choose; the clearer, also, is the consciousness that accompanies them. But neither this mobility nor this choice nor consequently this consciousness involves as a necessary condition the presence of a nervous system; the latter has only canalized in definite directions, and brought up to a higher degree of intensity, a rudimentary and vague activity, diffused throughout the mass of the organized substance. The lower we descend in the animal series, the more the nervous centers are simplified, and the more, too, they separate from each other, till finally the nervous elements disappear, merged in the mass of a less-differentiated organism. But it is the same with all the other apparatus, with all the other anatomical elements; and it would be as absurd to refuse consciousness to an animal because it has no brain as to declare it incapable of nourishing itself because it has no stomach. The truth is that the nervous system arises, like the other systems, from a division of labor. It does not create the function, it only brings it to a higher degree of intensity and precision by giving it the double form of reflex and voluntary activity. To accomplish a true reflex movement, a whole mechanism is necessary, set up in the spinal cord or the medulla. To choose voluntarily between several definite courses of action, cerebral centers are necessary, that is, crossways from which paths start, leading to motor mechanisms of diverse form but equal precision. But where nervous elements are not yet canalized, still less concentrated into a system, there is something from which, by a kind of splitting, both the reflex and the voluntary will arise, something which has neither the mechanical precision of the former nor the intelligent hesitations of the latter, but which, partaking of both, it may be infinitesimally, is a reaction simply undecided, and therefore vaguely conscious. This amounts to saying that the humblest organism is conscious in proportion to its power to move *freely*. Is consciousness here, in relation to movement, the effect or the cause? In one sense it is the cause, since it has to direct locomotion. But in another sense it is the effect, for it is the motor activity that maintains it, and, once this activity disappears, consciousness dies away or rather falls asleep. In crustaceans such as the rhizocephala, which must formerly have shown a more differentiated structure, fixity and parasitism accompany the degeneration and almost complete disappearance of the nervous system. Since, in such a case, the progress of organization must have localized all the conscious activity in nervous centers, we may conjecture that consciousness is even weaker in animals of this kind than in organisms much less differentiated, which have never had nervous centers but have remained mobile.

How then could the plant, which is fixed in the earth and finds its food on the spot, have developed in the direction of conscious activity? The membrane of cellulose, in which the protoplasm wraps itself up, not only prevents the simplest vegetable organism from moving, but screens it also, in some measure, from those outer stimuli which act on the sensibility of the animal as irritants and prevent it from going to sleep.[1] The plant is therefore unconscious. Here again, however, we must beware of radical distinctions. "Unconscious" and "conscious" are not two labels which can be mechanically fastened, the one on every vegetable cell, the other on all animals. While consciousness sleeps in the animal which has degenerated into a motionless parasite, it probably awakens in the vegetable that has regained liberty of movement, and awakens in just the degree to which the vegetable has reconquered this liberty. Nevertheless, consciousness and unconsciousness mark the directions in which the two kingdoms have developed, in this sense, that to find the best specimens of consciousness in the animal we must *ascend* to the highest representatives of the series, whereas, to find probable cases of vegetable consciousness, we must *descend* as low as possible in the scale of plants—down to the zoospores of the algae, for instance, and, more generally, to those unicellular organisms which may be said to hesitate between the vegetable form and animality. From this standpoint, and in this measure, we should define the animal by sensibility and awakened consciousness, the vegetable by consciousness asleep and by insensibility.

To sum up, the vegetable manufactures organic substances directly with mineral substances; as a rule, this aptitude enables it to dispense with movement and so with feeling. Animals, which are obliged to go in search of their food, have evolved in the direction of locomotor

[1] Cope, *op. cit.* p. 76.

activity, and consequently of a consciousness more and more distinct, more and more ample.

Now, it seems to us most probable that the animal cell and the vegetable cell are derived from a common stock, and that the first living organisms oscillated between the vegetable and animal form, participating in both at once. Indeed, we have just seen that the characteristic tendencies of the evolution of the two kingdoms, although divergent, coexist even now, both in the plant and in the animal. The proportion alone differs. Ordinarily, one of the two tendencies covers or crushes down the other, but in exceptional circumstances the suppressed one starts up and regains the place it had lost. The mobility and consciousness of the vegetable cell are not so sound asleep that they cannot rouse themselves when circumstances permit or demand it; and, on the other hand, the evolution of the animal kingdom has always been retarded, or stopped, or dragged back, by the tendency it has kept toward the vegetative life. However full, however overflowing the activity of an animal species may appear, torpor and unconsciousness are always lying in wait for it. It keeps up its rôle only by effort, at the price of fatigue. Along the route on which the animal has evolved, there have been numberless shortcomings and cases of decay, generally associated with parasitic habits; they are so many shuntings on to the vegetative life. Thus, everything bears out the belief that vegetable and animal are descended from a common ancestor which united the tendencies of both in a rudimentary state.

But the two tendencies mutually implied in this rudimentary form became dissociated as they grew. Hence the world of plants with its fixity and insensibility, hence the animals with their mobility and consciousness. There is no need, in order to explain this dividing into two, to bring in any mysterious force. It is enough to point out that the living being leans naturally toward what is most convenient to it, and that vegetables and animals have chosen two different kinds of convenience in the way of procuring the carbon and nitrogen they need. Vegetables continually and mechanically draw these elements from an environment that continually provides it. Animals, by action that is discontinuous, concentrated in certain moments, and conscious, go to find these bodies in organisms that have already fixed them. They are two different ways of being industrious, or perhaps we may prefer to say, of being idle. For this very reason we doubt whether nervous elements, however rudimentary, will ever be found in the plant. What corresponds in it to the directing will of the animal is, we believe, the direction in which it bends the energy of the solar radiation when it uses it to break the connection of the carbon with the oxygen in carbonic acid. What corresponds in it to the sensibility of the animal is the impressionability, quite of its kind, of its chlorophyl to light. Now, a nervous system being pre-eminently a mechanism which serves as intermediary between sensations and volitions, the true "nervous system" of the plant seems to be the mechanism or rather chemicism *sui generis* which serves as intermediary between the impressionability of its chlorophyl to light and the producing of starch: which amounts to saying that the plant can have no nervous elements, and that *the same impetus that has led the animal to give itself nerves and nerve centers must have ended, in the plant, in the chlorophyllian function.*[1]

This first glance over the organized world will enable us to ascertain more precisely what unites the two kingdoms, and also what separates them.

Suppose, as we suggested in the preceding chapter, that at the root of life there is an effort to engraft on to the necessity of physical forces the largest possible amount of *indetermination*. This effort cannot result in the creation of energy, or, if it does, the quantity created does not belong to the order of magnitude apprehended by our senses and instruments of measurement, our experience and science. All that the effort can do, then, is to make the best of a pre-existing energy which it finds at its disposal. Now, it finds only one way of succeeding in this, namely, to secure such an accumulation of potential energy from matter, that it can get, at any moment, the amount of work it needs for its action, simply by pulling a trigger. The effort itself possesses only that power of releasing. But the work of releasing, although always the same and always smaller than any given quantity, will be the more effective the heavier the weight it makes fall and the greater the height—or. in

[1] Just as the plant, in certain cases, recovers the faculty of moving actively which slumbers in it, so the animal, in exceptional circumstances, can replace itself in the conditions of the vegetative life and develop in itself an equivalent of the chlorophyllian function. It appears, indeed, from recent experiments of Maria von Linden, that the chrysalides and the caterpillars of certain lepidoptera, under the influence of light, fix the carbon of the carbonic acid contained in the atmosphere (M. von Linden, "L'Assimilation de l'acide carbonique par les chrysalides de Lépidoptères," *C. R. de la Soc. de biologie*, 1905, pp. 692 ff.).

other words, the greater the sum of potential energy accumulated and disposable. As a matter of fact, the principal source of energy usable on the surface of our planet is the sun. So the problem was this: to obtain from the sun that it should partially and provisionally suspend, here and there, on the surface of the earth, its continual outpour of usable energy, and store a certain quantity of it, in the form of unused energy, in appropriate reservoirs, whence it could be drawn at the desired moment, at the desired spot, in the desired direction. The substances forming the food of animals are just such reservoirs. Made of very complex molecules holding a considerable amount of chemical energy in the potential state, they are like explosives which only need a spark to set free the energy stored within them. Now, it is probable that life tended at the beginning to compass at one and the same time both the manufacture of the explosive and the explosion by which it is utilized. In this case, the same organism that had directly stored the energy of the solar radiation would have expended it in free movements in space. And for that reason we must presume that the first living beings sought on the one hand to accumulate, without ceasing, energy borrowed from the sun, and on the other hand to expend it, in a discontinuous and explosive way, in movements of locomotion. Even today, perhaps, a chlorophyl-bearing Infusorian such as the Euglena may symbolize this primordial tendency of life, though in a mean form, incapable of evolving. Is the divergent development of the two kingdoms related to what one may call the oblivion of each kingdom as regards one of the two halves of the program? Or rather, which is more likely, was the very nature of the matter, that life found confronting it on our planet, opposed to the possibility of the two tendencies evolving very far together in the same organism? What is certain is that the vegetable has trended principally in the first direction and the animal in the second. But if, from the very first, in making the explosive, nature had for object the explosion, then it is the evolution of the animal, rather than that of the vegetable, that indicates, on the whole, the fundamental direction of life.

The "harmony" of the two kingdoms, the complementary characters they display, might then be due to the fact that they develop two tendencies which at first were fused in one. The more the single original tendency grows, the harder it finds it to keep united in the same living being those two elements which in the rudimentary state implied each other. Hence a parting in two, hence two divergent evolutions; hence also two series of characters opposed in certain points, complementary in others, but, whether opposed or complementary, always preserving an appearance of kinship. While the animal evolved, not without accidents along the way, toward a freer and freer expenditure of discontinuous energy, the plant perfected rather its system of accumulation without moving. We shall not dwell on this second point. Suffice it to say that the plant must have been greatly benefited, in its turn, by a new division, analogous to that between plants and animals. While the primitive vegetable cell had to fix by itself both its carbon and its nitrogen, it became able almost to give up the second of these two functions as soon as microscopic vegetables came forward which leaned in this direction exclusively, and even specialized diversely in this still-complicated business.

The microbes that fix the nitrogen of the air and those which convert the ammoniacal compounds into nitrous ones, and these again into nitrates, have, by the same splitting up of a tendency primitively one, rendered to the whole vegetable world the same kind of service as the vegetables in general have rendered to animals. If a special kingdom were to be made for these microscopic vegetables, it might be said that in the microbes of the soil, the vegetables and the animals, we have before us the *analysis*, carried out by the matter that life found at its disposal on our planet, of all that life contained, at the outset, in a state of reciprocal implication. Is this, properly speaking, a "division of labor"? These words do not give the exact idea of evolution, such as we conceive it. Wherever there is division of labor, there is *association* and also *convergence* of effort. Now, the evolution we are speaking of is never achieved by means of association, but by *dissociation*; it never tends toward convergence, but toward *divergence* of efforts. The harmony between terms that are mutually complementary in certain points is not, in our opinion, produced, in course of progress, by a reciprocal adaptation; on the contrary, it is complete only at the start. It arises from an original identity, from the fact that the evolutionary process, splaying out like a sheaf, sunders, in proportion to their simultaneous growth, terms which at first completed each other so well that they coalesced.

Now, the elements into which a tendency splits up are far from possessing the same importance, or, above all, the same power to evolve. We have just distinguished three different kingdoms, if one may so express it, in the organized world. While the first comprises only microorganisms which have remained in the rudimentary state, animals and vegetables have taken their flight toward very lofty fortunes. Such, indeed, is generally the case when a tendency divides. Among the divergent developments to which it gives rise, some go on indefinitely, others come more or less quickly to the end of their tether. These latter do not issue directly from the primitive tendency, but from one of the elements into which it has divided; they are residual developments made and left behind on the way by some truly elementary tendency which continues to evolve. Now, these truly elementary tendencies, we think, bear a mark by which they may be recognized.

This mark is like a trace, still visible in each, of what was in the original tendency of which they represent the elementary directions. The elements of a tendency are not like objects set beside each other in space and mutually exclusive, but rather like psychic states, each of which, although it be itself to begin with, yet partakes of others, and so virtually includes in itself the whole personality to which it belongs. There is no real manifestation of life, we said, that does not show us, in a rudimentary or latent state, the characters of other manifestations. Conversely, when we meet, on one line of evolution, a recollection, so to speak, of what is developed along other lines, we must conclude that we have before us dissociated elements of one and the same original tendency. In this sense, vegetables and animals represent the two great divergent developments of life. Though the plant is distinguished from the animal by fixity and insensibility, movement and consciousness sleep in it as recollections which may waken. But, beside these normally sleeping recollections, there are others awake and active, just those, namely, whose activity does not obstruct the development of the elementary tendency itself. We may then formulate this law: *When a tendency splits up in the course of its development, each of the special tendencies which thus arise tries to preserve and develop everything in the primitive tendency that is not incompatible with the work for which it is specialized.* This explains precisely the fact we dwelt on in the preceding chapter, viz., the formation of identical complex mechanisms on independent lines of evolution. Certain deep-seated analogies between the animal and the vegetable have probably no other cause: sexual generation is perhaps only a luxury for the plant, but to the animal it was a necessity, and the plant must have been driven to it by the same impetus which impelled the animal thereto, a primitive, original impetus, anterior to the separation of the two kingdoms. The same may be said of the tendency of the vegetable toward a growing complexity. This tendency is essential to the animal kingdom, ever tormented by the need of more and more extended and effective action. But the vegetable, condemned to fixity and insensibility, exhibits the same tendency only because it received at the outset the same impulsion. Recent experiments show that it varies at random when the period of "mutation" arrives; whereas the animal must have evolved, we believe, in much more definite directions. But we will not dwell further on this original doubling of the modes of life. Let us come to the evolution of animals, in which we are more particularly interested.

What constitutes animality, we said, is the faculty of

utilizing a releasing mechanism for the conversion of as much stored-up potential energy as possible into "explosive" actions. In the beginning the explosion is haphazard, and does not choose its direction. Thus the amoeba thrusts out its pseudopodic prolongations in all directions at once. But, as we rise in the animal scale, the form of the body itself is observed to indicate a certain number of very definite directions along which the energy travels. These directions are marked by so many chains of nervous elements. Now, the nervous element has gradually emerged from the barely differentiated mass of organized tissue. It may, therefore, be surmised that in the nervous element, as soon as it appears, and also in its appendages, the faculty of suddenly freeing the gradually stored-up energy is concentrated. No doubt, every living cell expends energy without ceasing, in order to maintain its equilibrium. The vegetable cell, torpid from the start, is entirely absorbed in this work of maintenance alone, as if it took for end what must at first have been only a means. But, in the animal, all points to action, that is, to the utilization of energy for movements from place to place. True, every animal cell expends a good deal—often the whole—of the energy at its disposal in keeping itself alive; but the organism as a whole tries to attract as much energy as possible to those points where the locomotive movements are effected. So that where a nervous system exists, with its complementary sense-organs and motor apparatus, everything should happen as if the rest of the body had, as its essential function, to prepare for these and pass on to them, at the moment required, that force which they are to liberate by a sort of explosion.

The part played by food amongst the higher animals is, indeed, extremely complex. In the first place it serves to repair tissues, then it provides the animal with the heat necessary to render it as independent as possible of changes in external temperature. Thus it preserves, supports and maintains the organism in which the nervous system is set and on which the nervous elements have to live. But these nervous elements would have no reason for existence if the organism did not pass to them, and especially to the muscles they control, a certain energy to expend; and it may even be conjectured that there, in the main, is the essential and ultimate destination of food. This does not mean that the greater part of the food is used in this work. A state may have to make enormous expenditure to secure the return of taxes, and the sum which it will have to dispose of, after deducting the cost of collection, will perhaps be very small: that sum is, none the less, the reason for the tax and for all that has been spent to obtain its return. So it is with the energy which the animal demands of its food.

Many facts seem to indicate that the nervous and muscular elements stand in this relation toward the rest of the organism. Glance first at the distribution of alimentary substances among the different elements of the living body. These substances fall into two classes, one the quaternary or albuminoid, the other the ternary, including the carbohydrates and the fats. The albuminoids are properly plastic, destined to repair the tissues—although, owing to the carbon they contain, they are capable of providing energy on occasion. But the function of supplying energy has developed more particularly on the second class of substances: these, being deposited in the cell rather than forming part of its substance, convey to it, in the form of chemical potential, an expansive energy that may be directly converted into either movement or heat. In short, the chief function of the albuminoids is to repair the machine, while the function of the other class of substances is to supply power. It is natural that the albuminoids should have no specially allotted destination, since every part of the machine has to be maintained. But not so with the other substances. The carbohydrates are distributed very unequally, and this inequality of distribution seems to us in the highest degree instructive.

Conveyed by the arterial blood in the form of glucose, these substances are deposited, in the form of glycogen, in the different cells forming the tissues. We know that one of the principal functions of the liver is to maintain at a constant level the quantity of glucose held by the blood, by means of the reserves of glycogen secreted by the hepatic cells. Now, in this circulation of glucose and accumulation of glycogen, it is easy to see that the effect is as if the whole effort of the organism were directed toward providing with potential energy the elements of both the muscular and the nervous tissues. The organism proceeds differently in the two cases, but it arrives at the same result. In the first case, it provides the muscle-cell with a large reserve deposited in advance: the quantity of glycogen contained in the muscles is, indeed, enormous in comparison with what is found in the other tissues. In the nervous tissue, on the contrary, the reserve is small (the nervous elements, whose function is merely to liberate the potential energy stored in the muscle, never have to furnish much work at one time);

but the remarkable thing is that this reserve is restored by the blood at the very moment that it is expended, so that the nerve is instantly recharged with potential energy. Muscular tissue and nervous tissue are, therefore, both privileged, the one in that it is stocked with a large reserve of energy, the other in that it is always served at the instant it is in need and to the exact extent of its requirements.

More particularly, it is from the sensory-motor system that the call for glycogen, the potential energy, comes, as if the rest of the organism were simply there in order to transmit force to the nervous system and to the muscles which the nerves control. True, when we think of the part played by the nervous system (even the sensory-motor system) as regulator of the organic life, it may well be asked whether, in this exchange of good offices between it and the rest of the body, the nervous system is indeed a master that the body serves. But we shall already incline to this hypothesis when we consider, even in the static state only, the distribution of potential energy among the tissues; and we shall be entirely convinced of it when we reflect upon the conditions in which the energy is expended and restored. For suppose the sensory-motor system is a system like the others, of the same rank as the others. Borne by the whole of the organism, it will wait until an excess of chemical potential is supplied to it before it performs any work. In other words, it is the production of glycogen which will regulate the consumption by the nerves and muscles. On the contrary, if the sensory-motor system is the actual master, the duration and extent of its action will be independent, to a certain extent at least, of the reserve of glycogen that it holds, and even of that contained in the whole of the organism. It will perform work, and the other tissues will have to arrange as they can to supply it with potential energy. Now, this is precisely what does take place, as is shown in particular by the experiments of Morat and Dufourt.[1] While the glycogenic function of the liver depends on the action of the excitory nerves which control it, the action of these nerves is subordinated to the action of those which stimulate the locomotor muscles—in this sense, that the muscles begin by expending without calculation, thus consuming glycogen, impoverishing the blood of its glucose, and finally causing the liver, which has had to pour into the impoverished blood some of its reserve of glycogen, to manufacture a fresh supply. From the sensory-motor system, then, everything starts; on that system everything converges; and we may say, without metaphor, that the rest of the organism is at its service.

Consider again what happens in a prolonged fast. It is a remarkable fact that in animals that have died of hunger the brain is found to be almost unimpaired, while the other organs have lost more or less of their weight and their cells have undergone profound changes.[2] It seems as though the rest of the body had sustained the

[1] *Archives de physiologie*, 1892.

[2] De Manacéine, "Quelques observations expérimentales sur l'influence de l'insomnie absolue" (*Arch. ital. de biologie*, t. xxi., 1894, pp. 322 ff.). Recently, analogous observations have been made on a man who died of inanition after a fast of thirty-five days. See, on this subject, in the *Année biologique* of 1898, p. 338, the résumé of an article (in Russian) by Tarakevitch and Stchasny.

nervous system to the last extremity, treating itself simply as the means of which the nervous system is the end.

To sum up: if we agree, in short, to understand by "the sensory-motor system" the cerebro-spinal nervous system together with the sensorial apparatus in which it is prolonged and the locomotor muscles it controls, we may say that a higher organism is essentially a sensory-motor system installed on systems of digestion, respiration, circulation, secretion, etc., whose function it is to repair, cleanse and protect it, to create an unvarying internal environment for it, and above all to pass it potential energy to convert into locomotive movement.[1] It is true that the more the nervous function is perfected, the more must the functions required to maintain it develop, and the more exacting, consequently, they become for themselves. As the nervous activity has emerged from the protoplasmic mass in which it was almost drowned, it has had to summon around itself activities of all kinds for its support. These could only be developed on other activities, which again implied others, and so on indefinitely. Thus it is that the complexity of functioning of the higher organisms goes on to infinity. The study of

[1] Cuvier said: "The nervous system is, at bottom, the whole animal; the other systems are there only to serve it." ("Sur un nouveau rapprochement à établir entre les classes qui composent le regne animal," *Arch. du Muséum d'histoire naturelle*, Paris, 1812, pp. 73-84.) Of course, it would be necessary to apply a great many restrictions to this formula—for example, to allow for the cases of degradation and retrogression in which the nervous system passes into the background. And, moreover, with the nervous system must be included the sensorial apparatus on the one hand and the motor on the other, between which it acts as intermediary. Cf. Foster, art. "Physiology," in the *Encyclopaedia Britannica*, Edinburgh, 1885, p. 17.

one of these organisms therefore takes us round in a circle, as if everything was a means to everything else. But the circle has a center, none the less, and that is the system of nervous elements stretching between the sensory organs and the motor apparatus.

We will not dwell here on a point we have treated at length in a former work. Let us merely recall that the progress of the nervous system has been effected both in the direction of a more precise adaptation of movements and in that of a greater latitude left to the living being to choose between them. These two tendencies may appear antagonistic, and indeed they are so; but a nervous chain, even in its most rudimentary form, successfully reconciles them. On the one hand, it marks a well-defined track between one point of the periphery and another, the one sensory, the other motor. It has therefore canalized an activity which was originally diffused in the protoplasmic mass. But, on the other hand, the elements that compose it are probably discontinuous; at any rate, even supposing they anastomose, they exhibit a *functional* discontinuity, for each of them ends in a kind of crossroad where probably the nervous current may choose its course. From the humblest Monera to the best-endowed insects, and up to the most intelligent vertebrates, the progress realized has been above all a progress of the nervous system, coupled at every stage with all the new constructions and complications of mechanism that this progress required. As we foreshadowed in the beginning of this work, the rôle of life is to insert some *indetermination* into matter. Indeterminate, *i.e.* unforeseeable, are the forms it creates in the course of its evolution. More and more indeterminate also, more and more free, is the activity to which these forms serve as the vehicle. A nervous system, with neurones placed end to end in such wise that, at the extremity of each, manifold ways open in which manifold questions present themselves, is a veritable *reservoir of indetermination*. That the main energy of the vital impulse has been spent in creating apparatus of this kind is, we believe, what a glance over the organized world as a whole easily shows. But concerning the vital impulse itself a few explanations are necessary.

It must not be forgotten that the force which is evolving throughout the organized world is a limited force, which is always seeking to transcend itself and always remains inadequate to the work it would fain produce. The errors and puerilities of radical finalism are due to the misapprehension of this point. It has represented the whole of the living world as a construction, and a construction analogous to a human work. All the pieces have been arranged with a view to the best possible functioning of the machine. Each species has its reason for existence, its part to play, its allotted place; and all join together, as it were, in a musical concert, wherein the seeming discords are really meant to bring out a fundamental harmony. In short, all goes on in nature as in the works of human genius, where, though the result may be trifling, there is at least perfect adequacy between the object made and the work of making it.

Nothing of the kind in the evolution of life. There, the disproportion is striking between the work and the result. From the bottom to the top of the organized world we do indeed find one great effort; but most often this effort turns short, sometimes paralyzed by contrary forces, sometimes diverted from what it should do by what it does, absorbed by the form it is engaged in taking, hypnotized by it as by a mirror. Even in its most perfect works, though it seems to have triumphed over external resistances and also over its own, it is at the mercy of the materiality which it has had to assume. It is what each of us may experience in himself. Our freedom, in the very movements by which it is affirmed, creates the growing habits that will stifle it if it fails to renew itself by a constant effort: it is dogged by automatism. The most living thought becomes frigid in the formula that expresses it. The word turns against the idea.

The letter kills the spirit. And our most ardent enthusiasm, as soon as it is externalized into action, is so naturally congealed into the cold calculation of interest or vanity, the one takes so easily the shape of the other, that we might confuse them together, doubt our own sincerity, deny goodness and love, if we did not know that the dead retain for a time the features of the living.

The profound cause of this discordance lies in an irremediable difference of rhythm. Life in general is mobility itself; particular manifestations of life accept this mobility reluctantly, and constantly lag behind. It is always going ahead; they want to mark time. Evolution in general would fain go on in a straight line; each special evolution is a kind of circle. Like eddies of dust raised by the wind as it passes, the living turn upon themselves, borne up by the great blast of life. They are therefore relatively stable, and counterfeit immobility so well that we treat each of them as a *thing* rather than as a *progress*, forgetting that the very permanence of their form is only the outline of a movement. At times, however, in a fleeting vision, the invisible breath that bears them is materialized before our eyes. We have this sudden illumination before certain forms of maternal love, so striking, and in most animals so touching, observable even in the solicitude of the plant for its seed. This love, in which some have seen the great mystery of life, may possibly deliver us life's secret. It shows us each generation leaning over the generation that shall follow. It allows us a glimpse of the fact that the living being is above all a thoroughfare, and that the essence of life is in the movement by which life is transmitted.

This contrast between life in general, and the forms in which it is manifested, has everywhere the same character. It might be said that life tends toward the utmost possible action, but that each species prefers to contribute the slightest possible effort. Regarded in what constitutes its true essence, namely, as a transition from species to species, life is a continually growing action. But each of the species, through which life passes, aims only at its own convenience. It goes for that which demands the least labor. Absorbed in the form it is about to take, it falls into a partial sleep, in which it ignores almost all the rest of life; it fashions itself so as to take the greatest possible advantage of its immediate environment with the least possible trouble. Accordingly, the act by which life goes forward to the creation of a new form, and the act by which this form is shaped, are two different and often antagonistic movements. The first is continuous with the second, but cannot continue in it without being drawn aside from its direction, as would happen to a man leaping, if, in order to clear the obstacle, he had to turn his eyes from it and look at himself all the while.

Living forms are, by their very definition, forms that are able to live. In whatever way the adaptation of the organism to its circumstances is explained, it has necessarily been sufficient, since the species has subsisted. In this sense, each of the successive species that paleontology and zoology describe was a *success* carried off by life. But we get a very different impression when we refer each species to the movement that has left it behind on its way, instead of to the conditions into which it has been set. Often this movement has turned aside; very often, too, it has stopped short; what was to have been a thoroughfare has become a terminus. From this new point of view, failure seems the rule, success exceptional and always imperfect. We shall see that, of the four main directions along which animal life bent its course, two have led to blind alleys, and, in the other two, the effort has generally been out of proportion to the result.

Documents are lacking to reconstruct this history in detail, but we can make out its main lines. We have already said that animals and vegetables must have separated soon from their common stock, the vegetable falling asleep in immobility, the animal, on the contrary, becoming more and more awake and marching on to the conquest of a nervous system. Probably the effort of the animal kingdom resulted in creating organisms still very simple, but endowed with a certain freedom of action, and, above all, with a shape so undecided that it could lend itself to any future determination. These animals may have resembled some of our worms, but with this difference, however, that the worms living today, to which they could be compared, are but the empty and fixed examples of infinitely plastic forms, pregnant with an unlimited future, the common stock of the echinoderms, mollusks, arthropods and vertebrates.

One danger lay in wait for them, one obstacle which might have stopped the soaring course of animal life. There is one peculiarity with which we cannot help being struck when glancing over the fauna of primitive times, namely, the imprisonment of the animal in a more or less solid sheath, which must have obstructed and often even paralyzed its movements. The mollusks of that time had a shell more universally than those of today. The arthropods in general were provided with a carapace; most of them were crustaceans. The more ancient fishes had a bony sheath of extreme hardness.[1] The explanation of this general fact should be sought, we believe, in a tendency of soft organisms to defend themselves against one another by making themselves, as far as possible, undevourable. Each species, in the act by which it comes into being, trends toward that which is most expedient. Just as among primitive organisms there were some that turned toward animal life by refusing to manufacture organic out of inorganic material and taking organic substances ready made from organisms that had turned toward the vegetative life, so, among the animal species themselves, many contrived to live at the expense of other animals. For an organism that is animal, that is to say mobile, can avail itself of its mobility

[1] See, on these different points, the work of Gaudry, *Essai de paléontologie philosophique*, Paris, 1896, pp. 14-16 and 78-79.

to go in search of defenseless animals, and feed on them quite as well as on vegetables. So, the more species became mobile, the more they became voracious and dangerous to one another. Hence a sudden arrest of the entire animal world in its progress toward higher and higher mobility; for the hard and calcareous skin of the echinoderm, the shell of the mollusk, the carapace of the crustacean and the ganoid breast-plate of the ancient fishes probably all originated in a common effort of the animal species to protect themselves against hostile species. But this breast-plate, behind which the animal took shelter, constrained it in its movements and sometimes fixed it in one place. If the vegetable renounced consciousness in wrapping itself in a cellulose membrane, the animal that shut itself up in a citadel or in armor condemned itself to a partial slumber. In this torpor the echinoderms and even the mollusks live today. Probably arthropods and vertebrates were threatened with it too. They escaped, however, and to this fortunate circumstance is due the expansion of the highest forms of life.

In two directions, in fact, we see the impulse of life to movement getting the upper hand again. The fishes exchanged their ganoid breast-plate for scales. Long before that, the insects had appeared, also disencumbered of the breast-plate that had protected their ancestors. Both supplemented the insufficiency of their protective covering by an agility that enabled them to escape their enemies, and also to assume the offensive, to choose the place and the moment of encounter. We see a progress of the same kind in the evolution of human armaments. The first impulse is to seek shelter; the second, which is the better, is to become as supple as possible for flight and above all for attack—attack being the most effective means of defense. So the heavy hoplite was supplanted by the legionary; the knight, clad in armor, had to give place to the light free-moving infantryman; and in a general way, in the evolution of life, just as in the evolution of human societies and of individual destinies, the greatest successes have been for those who have accepted the heaviest risks.

Evidently, then, it was to the animal's interest to make itself more mobile. As we said when speaking of adaptation in general, any transformation of a species can be explained by its own particular interest. This will give the immediate cause of the variation, but often only the most superficial cause. The profound cause is the impulse which thrust life into the world, which made it divide into vegetables and animals, which shunted the animal on to suppleness of form, and which, at a certain moment, in the animal kingdom threatened with torpor, secured that, on some points at least, it should rouse itself up and move forward.

On the two paths along which the vertebrates and arthropods have separately evolved, development (apart from retrogressions connected with parasitism or any other cause) has consisted above all in the progress of the sensory-motor nervous system. Mobility and suppleness were sought for, and also—through many experimental attempts, and not without a tendency to excess of substance and brute force at the start—variety of movements. But this quest itself took place in divergent directions. A glance at the nervous system of the arthropods and that of the vertebrates shows us the difference.

In the arthropods, the body is formed of a series more or less long of rings set together; motor activity is thus distributed amongst a varying—sometimes a considerable—number of appendages, each of which has its special function. In the vertebrates, activity is concentrated in two pairs of members only, and these organs perform functions which depend much less strictly on their form.[1] The independence becomes complete in man, whose hand is capable of any kind of work.

That, at least, is what we see. But behind what is seen there is what may be surmised—two powers, immanent in life and originally intermingled, which were bound to part company in course of growth.

To define these powers, we must consider, in the evolution both of the arthropods and the vertebrates, the species which mark the culminating point of each. How is this point to be determined? Here again, to aim at geometrical precision will lead us astray. There is no single simple sign by which we can recognize that one species is more advanced than another on the same line of evolution. There are manifold characters, that must be compared and weighed in each particular case, in order to ascertain to what extent they are essential or accidental and how far they must be taken into account.

It is unquestionable, for example, that *success* is the most general criterion of superiority, the two terms being, up to a certain point, synonymous. By success must be understood, so far as the living being is concerned, an aptitude to develop in the most diverse environments, through the greatest possible variety of ob-

[1] See, on this subject, Shaler, *The Individual*, New York, 1900, pp. 118-125.

stacles, so as to cover the widest possible extent of ground. A species which claims the entire earth for its domain is truly a dominating and consequently superior species. Such is the human species, which represents the culminating point of the evolution of the vertebrates. But such also are, in the series of the articulate, the insects and in particular certain hymenoptera. It has been said of the ants that, as man is lord of the soil, they are lords of the sub-soil.

On the other hand, a group of species that has appeared late may be a group of degenerates; but, for that, some special cause of retrogression must have intervened. By right, this group should be superior to the group from which it is derived, since it would correspond to a more advanced stage of evolution. Now man is probably the latest comer of the vertebrates;[1] and in the insect series no species is later than the hymenoptera, unless it be the lepidoptera, which are probably degenerates, living parasitically on flowering plants.

So, by different ways, we are led to the same conclusion. The evolution of the arthropods reaches its culminating point in the insect, and in particular in the hymenoptera, as that of the vertebrates in man. Now, since instinct is nowhere so developed as in the insect world, and in no group of insects so marvelously as in the hy-

[1] This point is disputed by M. René Quinton, who regards the carnivorous and ruminant mammals, as well as certain birds, as subsequent to man (R. Quinton, *L'Eau de mer milieu organique*, Paris, 1904, p. 435). We may say here that our general conclusions, although very different from M. Quinton's, are not irreconcilable with them; for if evolution has really been such as we represent it, the vertebrates must have made an effort to maintain themselves in the most favorable conditions of activity—the very conditions, indeed, which life had chosen in the beginning.

menoptera, it may be said that the whole evolution of the animal kingdom, apart from retrogressions toward vegetative life, has taken place on two divergent paths, one of which led to instinct and the other to intelligence.

Vegetative torpor, instinct and intelligence—these, then, are the elements that coincided in the vital impulsion common to plants and animals, and which, in the course of a development in which they were made manifest in the most unforeseen forms, have been dissociated by the very fact of their growth. *The cardinal error which, from Aristotle onwards, has vitiated most of the philosophies of nature, is to see in vegetative, instinctive and rational life, three successive degrees of the development of one and the same tendency, whereas they are three divergent directions of an activity that has split up as it grew.* The difference between them is not a difference of intensity, nor, more generally, of degree, but of kind.

It is important to investigate this point. We have seen in the case of vegetable and animal life how they are at once mutually complementary and mutually antagonistic. Now we must show that intelligence and instinct also are opposite and complementary. But let us first explain why we are generally led to regard them as activities of which one is superior to the other and based upon it, whereas in reality they are not things of the same order: they have not succeeded one another, nor can we assign to them different grades.

It is because intelligence and instinct, having originally been interpenetrating, retain something of their common origin. Neither is ever found in a pure state.

We said that in the plant the consciousness and mobility of the animal, which lie dormant, can be awakened; and that the animal lives under the constant menace of being drawn aside to the vegetative life. The two tendencies —that of the plant and that of the animal—were so thoroughly interpenetrating, to begin with, that there has never been a complete severance between them: they haunt each other continually; everywhere we find them mingled; it is the proportion that differs. So with intelligence and instinct. There is no intelligence in which some traces of instinct are not to be discovered, more especially no instinct that is not surrounded with a fringe of intelligence. It is this fringe of intelligence that has been the cause of so many misunderstandings. From the fact that instinct is always more or less intelligent, it has been concluded that instinct and intelligence are things of the same kind, that there is only a difference of complexity or perfection between them, and, above all, that one of the two is expressible in terms of the other. In reality, they accompany each other only because they are complementary, and they are complementary only because they are different, what is instinctive in instinct being opposite to what is intelligent in intelligence.

We are bound to dwell on this point. It is one of the utmost importance.

Let us say at the outset that the distinctions we are going to make will be too sharply drawn, just because we wish to define in instinct what is instinctive, and in intelligence what is intelligent, whereas all concrete instinct is mingled with intelligence, as all real intelligence is penetrated by instinct. Moreover, neither intelligence nor instinct lends itself to rigid definition: they are tendencies, and not things. Also, it must not be forgotten that in the present chapter we are considering intelligence and instinct as going out of life which deposits them along its course. Now the life manifested by an organism is, in our view, a certain effort to obtain certain things from the material world. No wonder, therefore, if it is the diversity of this effort that strikes us in instinct and intelligence, and if we see in these two modes of psychical activity, above all else, two different methods of action on inert matter. This rather narrow view of them has the advantage of giving us an objective means of distinguishing them. In return, however, it gives us, of intelligence in general and of instinct in general, only the mean position above and below which both constantly oscillate. For that reason the reader must expect to see in what follows only a diagrammatic drawing, in which the respective outlines of intelligence and instinct are sharper than they should be, and in which the shading-off which comes from the indecision of each and from their reciprocal encroachment on one another is neglected. In a matter so obscure, we cannot strive too hard for clearness. It will always be easy afterwards to soften the outlines and to correct what is too geometrical in the drawing—in short, to replace the rigidity of a diagram by the suppleness of life.

To what date is it agreed to ascribe the appearance of man on the earth? To the period when the first weapons, the first tools, were made. The memorable quarrel over the discovery of Boucher de Perthes in the quarry of

Moulin-Quignon is not forgotten. The question was whether real hatchets had been found or merely bits of flint accidentally broken. But that, supposing they were hatchets, we were indeed in the presence of intelligence, and more particularly of *human* intelligence, no one doubted for an instant. Now let us open a collection of anecdotes on the intelligence of animals: we shall see that besides many acts explicable by imitation or by the automatic association of images, there are some that we do not hesitate to call intelligent: foremost among them are those that bear witness to some idea of manufacture, whether the animal life succeeds in fashioning a crude instrument or uses for its profit an object made by man. The animals that rank immediately after man in the matter of intelligence, the apes and elephants, are those that can use an artificial instrument occasionally. Below, but not very far from them, come those that *recognize* a constructed object: for example, the fox, which knows quite well that a trap is a trap. No doubt, there is intelligence wherever there is inference; but inference, which consists in an inflection of past experience in the direction of present experience, is already a beginning of invention. Invention becomes complete when it is materialized in a manufactured instrument. Toward that achievement the intelligence of animals tends as toward an ideal. And though, ordinarily, it does not yet succeed in fashioning artificial objects and in making use of them, it is preparing for this by the very variations which it performs on the instincts furnished by nature. As regards human intelligence, it has not been sufficiently noted that mechanical invention has been from the first its essential feature, that even today our social life gravitates around the manufacture and use of artificial instruments, that the inventions which strew the road of progress have also traced its direction. This we hardly realize, because it takes us longer to change ourselves than to change our tools. Our individual and even social habits survive a good while the circumstances for which they were made, so that the ultimate effects of an invention are not observed until its novelty is already out of sight. A century has elapsed since the invention of the steam engine, and we are only just beginning to feel the depths of the shock it gave us. But the revolution it has effected in industry has nevertheless upset human relations altogether. New ideas are arising, new feelings are on the way to flower. In thousands of years, when, seen from the distance, only the broad lines of the present age will still be visible, our wars and our revolutions will count for little, even supposing they are remembered at all; but the steam engine, and the procession of inventions of every kind that accompanied it, will perhaps be spoken of as we speak of the bronze or of the chipped stone of pre-historic times: it will serve to define an age.[1] If we could rid ourselves of all pride, if, to define our species, we kept strictly to what the historic and the prehistoric periods show us to be the constant characteristic of man and of intelligence, we should say not *Homo sapiens*, but *Homo faber*. In short, *intelligence, considered in what seems to be its original feature, is the faculty of manufacturing artificial objects, espe-*

[1] M. Paul Lacombe has laid great stress on the important influence that great inventions have exercised on the evolution of humanity (P. Lacombe, *De l'histoire considérée comme science*, Paris, 1894. See, in particular, pp. 168-247).

cially tools to make tools, and of indefinitely varying the manufacture.

Now, does an unintelligent animal also possess tools or machines? Yes, certainly, but here the instrument forms a part of the body that uses it; and, corresponding to this instrument, there is an *instinct* that knows how to use it. True, it cannot be maintained that *all* instincts consist in a natural ability to use an inborn mechanism. Such a definition would not apply to the instincts which Romanes called "secondary"; and more than one "primary" instinct would not come under it. But this definition, like that which we have provisionally given of intelligence, determines at least the ideal limit toward which the very numerous forms of instinct are traveling. Indeed, it has often been pointed out that most instincts are only the continuance, or rather the consummation, of the work of organization itself. Where does the activity of instinct begin? And where does that of nature end? We cannot tell. In the metamorphoses of the larva into the nymph and into the perfect insect, metamorphoses that often require appropriate action and a kind of initiative on the part of the larva, there is no sharp line of demarcation between the instinct of the animal and the organizing work of living matter. We may say, as we will, either that instinct organizes the instruments it is about to use, or that the process of organization is continued in the instinct that has to use the organ. The most marvelous instincts of the insect do nothing but develop its special structure into movements: indeed, where social life divides the labor among different individuals, and thus allots them different instincts, a corresponding difference of structure is observed: the polymorphism of ants, bees, wasps and certain pseudo-neuroptera is well known. Thus, if we consider only those typical cases in which the complete triumph of intelligence and of instinct is seen, we find this essential difference between them: *instinct perfected is a faculty of using and even of constructing organized instruments; intelligence perfected is the faculty of making and using unorganized instruments.*

The advantages and drawbacks of these two modes of activity are obvious. Instinct finds the appropriate instrument at hand: this instrument, which makes and repairs itself, which presents, like all the works of nature, an infinite complexity of detail combined with a marvelous simplicity of function, does at once, when required, what it is called upon to do, without difficulty and with a perfection that is often wonderful. In return, it retains an almost invariable structure, since a modification of it involves a modification of the species. Instinct is therefore necessarily specialized, being nothing but the utilization of a specific instrument for a specific object. The instrument constructed intelligently, on the contrary, is an imperfect instrument. It costs an effort. It is generally troublesome to handle. But, as it is made of unorganized matter, it can take any form whatsoever, serve any purpose, free the living being from every new difficulty that arises and bestow on it an unlimited number of powers. Whilst it is inferior to the natural instrument for the satisfaction of immediate wants, its advantage over it is the greater, the less urgent the need. Above all, it reacts on the nature of the being that constructs it; for in calling on him to exercise a new function, it confers on him, so to speak, a richer organization, being an artificial organ by which the natural organism is extended. For every need that it satisfies, it creates a new need; and so, instead of closing, like instinct, the round of action within which the animal tends to move automatically, it lays open to activity an unlimited field into which it is driven further and further, and made more and more free. But this advantage of intelligence over instinct only appears at a late stage, when intelligence, having raised construction to a higher degree, proceeds to construct constructive machinery. At the outset, the advantages and drawbacks of the artificial instrument and of the natural instrument balance so well that it is hard to foretell which of the two will secure to the living being the greater empire over nature.

We may surmise that they began by being implied in each other, that the original psychical activity included both at once, and that, if we went far enough back into the past, we should find instincts more nearly approaching intelligence than those of our insects, intelligence nearer to instinct than that of our vertebrates, intelligence and instinct being, in this elementary condition, prisoners of a matter which they are not yet able to control. If the force immanent in life were an unlimited force, it might perhaps have developed instinct and intelligence together, and to any extent, in the same organisms. But everything seems to indicate that this force is limited, and that it soon exhausts itself in its very manifestation. It is hard for it to go far in several directions at once: it must choose. Now, it has the choice between two modes of acting on the material world: it can either effect this action *directly* by creating an *organized* instrument to work with; or else it can effect it *indirectly* through an organism which, instead of possessing the required instrument naturally, will itself construct it by fashioning inorganic matter. Hence intelligence and instinct, which diverge more and more as they develop, but which never entirely separate from each other. On the one hand, the most perfect instinct of the insect is accompanied by gleams of intelligence, if only in the choice of place, time and materials of construction: the bees, for example, when by exception they build in the open air, invent new and really intelligent arrangements to adapt themselves to such new conditions.[1] But, on the other hand, intelligence has even more need of instinct than instinct has of intelligence; for the power to give shape to crude matter involves already a superior degree of organization, a degree to which the animal could not have risen, save on the wings of instinct. So, while nature has frankly evolved in the direction of instinct in the arthropods, we observe in almost all the vertebrates the striving after rather than the expansion of intelligence. It is instinct still which forms the basis of their psychical activity; but intelligence is there, and would fain supersede it. Intelligence does not yet succeed in inventing instruments; but at least it tries to, by performing as many variations as possible on the instinct which it would like to dispense with. It gains complete self-possession only in man, and this triumph is attested by the very insufficiency of the natural means at man's disposal for defense against his enemies, against cold and hunger. This insufficiency, when we strive to fathom its significance, acquires the value of a prehistoric docu-

[1] Bouvier, "La Nidification des abeilles à l'air libre" (*C. R. de l'Ac. des sciences*, 7 mai 1906).

ment; it is the final leave-taking between intelligence and instinct. But it is no less true that nature must have hesitated between two modes of psychical activity—one assured of immediate success, but limited in its effects; the other hazardous, but whose conquests, if it should reach independence, might be extended indefinitely. Here again, then, the greatest success was achieved on the side of the greatest risk. *Instinct and intelligence therefore represent two divergent solutions, equally fitting, of one and the same problem.*

There ensue, it is true, profound differences of internal structure between instinct and intelligence. We shall dwell only on those that concern our present study. Let us say, then, that instinct and intelligence imply two radically different kinds of knowledge. But some explanations are first of all necessary on the subject of consciousness in general.

It has been asked how far instinct is conscious. Our reply is that there are a vast number of differences and degrees, that instinct is more or less conscious in certain cases, unconscious in others. The plant, as we shall see, has instincts; it is not likely that these are accompanied by feeling. Even in the animal there is hardly any complex instinct that is not unconscious in some part at least of its exercise. But here we must point out a difference, not often noticed, between two kinds of unconsciousness, viz., that in which consciousness is *absent*, and that in which consciousness is *nullified*. Both are equal to zero, but in one case the zero expresses the fact that there is nothing, in the other that we have two equal quantities of opposite sign which compensate and neutralize each other. The unconsciousness of a falling stone is of the former kind: the stone has no feeling of its fall. Is it the same with the unconsciousness of instinct, in the extreme cases in which instinct is unconscious? When we mechanically perform an habitual action, when the somnambulist automatically acts his dream, unconsciousness may be absolute; but this is merely due to the fact that the representation of the act is held in check by the performance of the act itself, which resembles the idea so perfectly, and fits it so exactly, that consciousness is unable to find room between them. *Representation is stopped up by action.* The proof of this is, that if the accomplishment of the act is arrested or thwarted by an obstacle, consciousness may reappear. It was there, but neutralized by the action which fulfilled and thereby filled the representation. The obstacle creates nothing positive; it simply makes a void, removes a stopper. This inadequacy of act to representation is precisely what we here call consciousness.

If we examine this point more closely, we shall find that consciousness is the light that plays around the zone of possible actions or potential activity which surrounds the action really performed by the living being. It signifies hesitation or choice. Where many equally possible actions are indicated without there being any real action (as in a deliberation that has not come to an end), consciousness is intense. Where the action performed is the only action possible (as in activity of the somnambulistic or more generally automatic kind), consciousness is reduced to nothing. Representation and knowledge exist none the less in the case if we find a whole series of systematized movements the last of which is already prefigured in the first, and if, besides, consciousness can flash out of them at the shock of an obstacle. From this point of view, *the consciousness of a living being may be defined as an arithmetical difference between potential and real activity. It measures the interval between representation and action.*

It may be inferred from this that intelligence is likely to point toward consciousness, and instinct toward unconsciousness. For, where the implement to be used is organized by nature, the material furnished by nature, and the result to be obtained willed by nature, there is little left to choice; the consciousness inherent in the representation is therefore counterbalanced, whenever it tends to disengage itself, by the performance of the act, identical with the representation, which forms its counterweight. Where consciousness appears, it does not so much light up the instinct itself as the thwartings to which instinct is subject; it is the *deficit* of instinct, the distance, between the act and the idea, that becomes consciousness so that consciousness, here, is only an accident. Essentially, consciousness only emphasizes the starting-point of instinct, the point at which the whole series of automatic movements is released. Deficit, on the contrary, is the normal state of intelligence. Laboring under difficulties is its very essence. Its original function being to construct unorganized instruments, it must, in spite of numberless difficulties, choose for this work the place and the time, the form and the matter. And it can never satisfy itself entirely, because every new satisfaction creates new needs. In short, while instinct and intelligence both involve knowledge, this knowledge is rather *acted* and unconscious in the case of instinct, *thought* and conscious in the case of intelligence. But it is a difference rather of degree than of kind. So long as consciousness is all we are concerned with, we close our eyes to what is, from the psychological point of view, the cardinal difference between instinct and intelligence.

In order to get at this essential difference we must, without stopping at the more or less brilliant light which illumines these two modes of internal activity, go straight to the two *objects*, profoundly different from each other, upon which instinct and intelligence are directed.

When the horse-fly lays its eggs on the legs or shoulders of the horse, it acts as if it knew that its larva has to develop in the horse's stomach and that the horse, in licking itself, will convey the larva into its digestive tract. When a paralyzing wasp stings its victim on just those points where the nervous centers lie, so as to render it motionless without killing it, it acts like a learned entomologist and a skilful surgeon rolled into one. But what shall we say of the little beetle, the Sitaris, whose story is so often quoted? This insect lays its eggs at the entrance of the underground passages dug by a kind of bee, the Anthophora. Its larva, after long waiting, springs upon the male Anthophora as it goes out of the passage, clings to it, and remains attached until the "nuptial flight," when it seizes the opportunity to pass from the male to the female, and quietly waits until it lays its eggs. It then leaps on the egg, which serves as a support for it in the honey, devours the egg in a few days, and, resting on the shell, undergoes its first metamorphosis. Organized now to float on the honey, it consumes this provision of nourishment, and becomes a nymph, then a perfect insect. Everything happens *as if* the larva of the

Sitaris, from the moment it was hatched, knew that the male Anthophora would first emerge from the passage; that the nuptial flight would give it the means of conveying itself to the female, who would take it to a store of honey sufficient to feed it after its transformation; that, until this transformation, it could gradually eat the egg of the Anthophora, in such a way that it could at the same time feed itself, maintain itself at the surface of the honey, and also suppress the rival that otherwise would have come out of the egg. And equally all this happens *as if* the Sitaris itself knew that its larva would know all these things. The knowledge, if knowledge there be, is only implicit. It is reflected outwardly in exact movements instead of being reflected inwardly in consciousness. It is none the less true that the behavior of the insect involves, or rather evolves, the idea of definite things existing or being produced in definite points of space and time, which the insect knows without having learned them.

Now, if we look at intelligence from the same point of view, we find that it also knows certain things without having learned them. But the knowledge in the two cases is of a very different order. We must be careful here not to revive again the old philosophical dispute on the subject of innate ideas. So we will confine ourselves to the point on which every one is agreed, to wit, that the young child understands immediately things that the animal will never understand, and that in this sense intelligence, like instinct, is an inherited function, therefore an innate one. But this innate intelligence, although it is a faculty of knowing, knows no object in particular. When the new-born babe seeks for the first time its mother's breast, so showing that it has knowledge (unconscious, no doubt) of a thing it has never seen, we say, just because the innate knowledge is in this case of a definite object, that it belongs to *instinct* and not to *intelligence*. Intelligence does not then imply the innate knowledge of any object. And yet, if intelligence knows nothing by nature, it has nothing innate. What, then, if it be ignorant of all things, can it know? Besides *things*, there are *relations*. The new-born child, so far as intelligent, knows neither definite objects nor a definite property of any object; but when, a little later on, he will hear an epithet being applied to a substantive, he will immediately understand what it means. The relation of attribute to subject is therefore seized by him naturally, and the same might be said of the general relation expressed by the verb, a relation so immediately conceived by the mind that language can leave it to be understood, as is instanced in rudimentary languages which have no verb. Intelligence, therefore, naturally makes use of relations of like with like, of content to container, of cause to effect, etc., which are implied in every phrase in which there is a subject, an attribute and a verb, expressed or understood. May one say that it has *innate* knowledge of each of these relations in particular? It is for logicians to discover whether they are so many irreducible relations, or whether they can be resolved into relations still more general. But, in whatever way we make the analysis of thought, we always end with one or several general categories, of which the mind possesses innate knowledge since it makes a natural use of them. Let us say, therefore, that *whatever, in instinct and intelligence, is innate knowledge, bears in the first case on* things *and in the second on* relations.

Philosophers distinguish between the matter of our knowledge and its form. The matter is what is given by the perceptive faculties taken in the elementary state. The form is the totality of the relations set up between these materials in order to constitute a systematic knowledge. Can the form, without matter, be an object of knowledge? Yes, without doubt, provided that this knowledge is not like a thing we possess so much as like a habit we have contracted—a direction rather than a state: it is, if we will, a certain natural bent of attention. The schoolboy, who knows that the master is going to dictate a fraction to him, draws a line before he knows what numerator and what denominator are to come; he therefore has present to his mind the general relation between the two terms although he does not know either of them; he knows the form without the matter. So is it, prior to experience, with the categories into which our experience comes to be inserted. Let us adopt then words sanctioned by usage, and give the distinction between intelligence and instinct this more precise formula: *Intelligence, in sc far as it is innate, is the knowledge of a* form; *instinct implies the knowledge of a* matter.

From this second point of view, which is that of knowledge instead of action, the force immanent in life in general appears to us again as a limited principle, in which originally two different and even divergent modes of knowing coexisted and intermingled. The first gets at definite objects immediately, in their materiality itself. It says, "This is what is." The second gets at no object in particular; it is only a natural power of relating an object to an object, or a part to a part, or an aspect to an aspect—in short, of drawing conclusions when in possession of the premises, of proceeding from what has been learnt to what is still unknown. It does not say, "This *is*"; it says only that "*if* the conditions are such, such will be the conditioned." In short, the first kind of knowledge, the instinctive, would be formulated in what philosophers call *categorical* propositions, while the second kind, the intellectual, would always be expressed *hypothetically*. Of these two faculties, the former seems, at first, much preferable to the other. And it would be so, in truth, if it extended to an endless number of objects. But, in fact, it applies only to one special object, and indeed only to a restricted part of that object. Of this, at least, its knowledge is intimate and full; not explicit, but implied in the accomplished action. The intellectual faculty, on the contrary, possesses naturally only an external and empty knowledge; but it has thereby the advantage of supplying a frame in which an infinity of objects may find room in turn. It is as if the force evolving in living forms, being a limited force, had had to choose between two kinds of limitation in the field of natural or innate knowledge, one applying to the *extension* of knowledge, the other to its *intension*. In the first case, the knowledge may be packed and full, but it will then be confined to one specific object; in the second, it is no longer limited by its object, but that is because it contains nothing, being only a form without matter. The two tendencies, at first implied in each other, had to separate in order to grow. They both went to seek their fortune in the world, and turned out to be instinct and intelligence.

Such, then, are the two divergent modes of knowledge by which intelligence and instinct must be defined, from the standpoint of knowledge rather than that of action. But knowledge and action are here only two aspects of one and the same faculty. It is easy to see, indeed, that the second definition is only a new form of the first.

If instinct is, above all, the faculty of using an organized natural instrument, it must involve innate knowledge (potential or unconscious, it is true), both of this instrument and of the object to which it is applied. Instinct is therefore innate knowledge of a *thing*. But intelligence is the faculty of constructing unorganized—that is to say artificial—instruments. If, on its account, nature gives up endowing the living being with the instruments that may serve him, it is in order that the living being may be able to vary his construction according to circumstances. The essential function of intelligence is therefore to see the way out of a difficulty in any circumstances whatever, to find what is most suitable, what answers best the question asked. Hence it bears essentially on the relations between a given situation and the means of utilizing it. What is innate in intellect, therefore, is the tendency to establish relations, and this tendency implies the natural knowledge of certain very general relations, a kind of stuff that the activity of each particular intellect will cut up into more special relations. Where activity is directed toward manufacture, therefore, knowledge necessarily bears on relations. But this entirely *formal* knowledge of intelligence has an immense advantage over the *material* knowledge of instinct. A form, just because it is empty, may be filled at will with any number of things in turn, even with those that are of no use. So that a formal knowledge is not limited to what is practically useful, although it is in view of practical utility that it has made its appearance in the world. An intelligent being bears within himself the means to transcend his own nature.

He transcends himself, however, less than he wishes, less also than he imagines himself to do. The purely formal character of intelligence deprives it of the ballast necessary to enable it to settle itself on the objects that are of the most powerful interest to speculation. Instinct, on the contrary, has the desired materiality, but it is incapable of going so far in quest of its object; it does not speculate. Here we reach the point that most concerns our present inquiry. The difference that we shall now proceed to denote between instinct and intelligence is what the whole of this analysis was meant to bring out. We formulate it thus: *There are things that intelligence alone is able to seek, but which, by itself, it will never find. These things instinct alone could find; but it will never seek them.*

It is necessary here to consider some preliminary details that concern the mechanism of intelligence. We have said that the function of intelligence is to establish relations. Let us determine more precisely the nature of these relations. On this point we are bound to be either vague or arbitrary so long as we see in the intellect a faculty intended for pure speculation. We are then reduced to taking the general frames of the understanding for something absolute, irreducible and inexplicable. The understanding must have fallen from heaven with its form, as each of us is born with his face. This form may be defined, of course, but that is all; there is no asking why it is what it is rather than anything else. Thus, it will be said that the function of the intellect is essentially unification, that the common object of all its operations is to introduce a certain unity into the diversity of phenomena, and so forth. But, in the first place, "unification" is a vague term, less clear than "relation" or even "thought," and says nothing more. And, moreover, it might be asked if the function of intelligence is not to divide even more than to unite. Finally, if the intellect proceeds as it does because it wishes to unite, and if it seeks unification simply because it has need of unifying, the whole of our knowledge becomes relative to certain requirements of the mind that probably might have been entirely different from what they are: for an intellect differently shaped, knowledge would have been different. Intellect being no longer dependent on anything, everything becomes dependent on it; and so, having placed the understanding too high, we end by putting too low the knowledge it gives us. Knowledge becomes relative, as soon as the intellect is made a kind of absolute.—We regard the human intellect, on the contrary, as relative to the needs of action. Postulate action, and the very form of the intellect can be deduced from it. This form is therefore neither irreducible nor inexplicable. And, precisely because it is not independent, knowledge cannot be said to depend on it: knowledge ceases to be a product of the intellect and becomes, in a certain sense, part and parcel of reality.

Philosophers will reply that action takes place in an *ordered* world, that this order is itself thought, and that we beg the question when we explain the intellect by action, which presupposes it. They would be right if our point of view in the present chapter was to be our final one. We should then be dupes of an illusion like that of

Spencer, who believed that the intellect is sufficiently explained as the impression left on us by the general characters of matter: as if the order inherent in matter were not intelligence itself! But we reserve for the next chapter the question up to what point and with what method philosophy can attempt a real genesis of the intellect at the same time as of matter. For the moment, the problem that engages our attention is of a psychological order. We are asking what is the portion of the material world to which our intellect is specially adapted. To reply to this question, there is no need to choose a system of philosophy: it is enough to take up the point of view of common sense.

Let us start, then, from action, and lay down that the intellect aims, first of all, at constructing. This fabrication is exercised exclusively on inert matter, in this sense, that even if it makes use of organized material, it treats it as inert, without troubling about the life which animated it. And of inert matter itself, fabrication deals only with the solid; the rest escapes by its very fluidity. If, therefore, the tendency of the intellect is to fabricate, we may expect to find that whatever is fluid in the real will escape it in part, and whatever is life in the living will escape it altogether. *Our intelligence, as it leaves the hands of nature, has for its chief object the unorganized solid.*

When we pass in review the intellectual functions, we see that the intellect is never quite at its ease, never entirely at home, except when it is working upon inert matter, more particularly upon solids. What is the most general property of the material world? It is extended: it presents to us objects external to other objects, and, in these objects, parts external to parts. Nc doubt, it is useful to us, in view of our ulterior manipulation, to regard each object as divisible into parts arbitrarily cut up, each part being again divisible as we like, and so on *ad infinitum*. But it is above all necessary, for our present manipulation, to regard the real object in hand, or the real elements into which we have resolved it, as *provisionally final*, and to treat them as so many *units*. To this possibility of decomposing matter as much as we please, and in any way we please, we allude when we speak of the *continuity* of material extension; but this continuity, as we see it, is nothing else but our ability, an ability that matter allows to us to choose the mode of discontinuity we shall find in it. It is always, in fact, the mode of discontinuity once chosen that appears to us as the actually real one and that which fixes our attention, just because it rules our action. Thus discontinuity is thought for itself; it is thinkable in itself; we form an idea of it by a positive act of our mind; while the intellectual representation of continuity is negative, being, at bottom, only the refusal of our mind, before any actually given system of decomposition, to regard it as the only possible one. *Of the discontinuous alone does the intellect form a clear idea.*

On the other hand, the objects we act on are certainly mobile objects, but the important thing for us to know is *whither* the mobile object is going and *where* it is at any moment of its passage. In other words, our interest is directed, before all, to its actual or future positions, and not to the *progress* by which it passes from one position to another, progress which is the movement itself. In our actions, which are systematized movements, what we fix our mind on is the end or meaning of the movement, its design as a whole—in a word, the immobile plan of its execution. That which really moves in action interests us only so far as the whole can be advanced, retarded, or stopped by any incident that may happen on the way. From mobility itself our intellect turns aside, because it has nothing to gain in dealing with it. If the intellect were meant for pure theorizing, it would take its place within movement, for movement is reality itself, and immobility is always only apparent or relative. But the intellect is meant for something altogether different. Unless it does violence to itself, it takes the opposite course; it always starts from immobility, as if this were the ultimate reality: when it tries to form an idea of movement, it does so by constructing movement out of immobilities put together. This operation, whose illegitimacy and danger in the field of speculation we shall show later on (it leads to deadlocks, and creates artificially insoluble philosophical problems), is easily justified when we refer it to its proper goal. Intelligence, in its natural state, aims at a practically useful end. When it substitutes for movement immobilities put together, it does not pretend to reconstitute the movement such as it actually is; it merely replaces it with a practical equivalent. It is the philosophers who are mistaken when they import into the domain of speculation a method of thinking which is made for action. But of this more anon. Suffice it now to say that to the stable and unchangeable our intellect is attached by virtue of its natural disposition. *Of immobility alone does the intellect form a clear idea.*

Now, fabricating consists in carving out the form of an object in matter. What is the most important is the form to be obtained. As to the matter, we choose that which is most convenient; but, in order to choose it, that is to say, in order to go and seek it among many others, we must have tried, in imagination at least, to endow every kind of matter with the form of the object conceived. In other words, an intelligence which aims at fabricating is an intelligence which never stops at the actual form of things nor regards it as final, but, on the contrary, looks upon all matter as if it were carvable at will. Plato compares the good dialectician to the skilful cook who carves the animal without breaking its bones, by following the articulations marked out by nature.[1] An intelligence which always proceeded thus would really be an intelligence turned toward speculation. But action, and in particular fabrication, requires the opposite mental tendency: it makes us consider every actual form of things, even the form of natural things, as artificial and provisional; it makes our thought effaee from the object perceived, even though organized and living, the lines that outwardly mark its inward structure; in short, it makes us regard its matter as indifferent to its form. The whole of matter is made to appear to our thought as an immense piece of cloth in which we can cut out what we will and sew it together again as we please. Let us note, in passing, that it is this power that we affirm when we say that there is a *space*, that is to say, a homogeneous and empty medium, infinite and infinitely divisible, lending itself indifferently to any mode of decomposition whatsoever. A medium of this kind is never perceived; it is only conceived. What is perceived is extension colored, resistant, divided according to the lines which mark out the boundaries of real bodies or of their real elements.

[1] Plato, *Thaedrus*, 265 E.

But when we think of our power over this matter, that is to say, of our faculty of decomposing and recomposing it as we please, we project the whole of these possible decompositions and recompositions behind real extension in the form of a homogeneous space, empty and indifferent, which is supposed to underlie it. This space is therefore, pre-eminently, the plan of our possible action on things, although, indeed, things have a natural tendency, as we shall explain further on, to enter into a frame of this kind. It is a view taken by mind. The animal has probably no idea of it, even when, like us, it perceives extended things. It is an idea that symbolizes the tendency of the human intellect to fabrication. But this point must not detain us now. Suffice it to say that *the intellect is characterized by the unlimited power of decomposing according to any law and of recomposing into any system.*

We have now enumerated a few of the essential features of human intelligence. But we have hitherto considered the individual in isolation, without taking account of social life. In reality, man is a being who lives in society. If it be true that the human intellect aims at fabrication, we must add that, for that as well as for other purposes, it is associated with other intellects. Now, it is difficult to imagine a society whose members do not communicate by signs. Insect societies probably have a language, and this language must be adapted, like that of man, to the necessities of life in common. By language community of action is made possible. But the requirements of joint action are not at all the same in a colony of ants and in a human society. In insect societies there is generally polymorphism, the subdivision of labor is natural, and each individual is riveted by its structure to the function it performs. In any case, these societies are based on instinct, and consequently on certain actions or fabrications that are more or less dependent on the form of the organs. So if the ants, for instance, have a language, the signs which compose it must be very limited in number, and each of them, once the species is formed, must remain invariably attached to a certain object or a certain operation: the sign is adherent to the thing signified. In human society, on the contrary, fabrication and action are of variable form, and, moreover, each individual must learn his part, because he is not preordained to it by his structure. So a language is required which makes it possible to be always passing from what is known to what is yet to be known. There must be a language whose signs—which cannot be infinite in number—are extensible to an infinity of things. This tendency of the sign to transfer itself from one object to another is characteristic of human language. It is observable in the little child as soon as he begins to speak. Immediately and naturally he extends the meaning of the words he learns, availing himself of the most accidental connection or the most distant analogy to detach and transfer elsewhere the sign that had been associated in his hearing with a particular object. "Anything can designate anything"; such is the latent principle of infantine language. This tendency has been wrongly confused with the faculty of generalizing. The animals themselves generalize; and, moreover, a sign—even an instinctive sign—always to some degree represents a genus. But what characterizes the signs of human language is not so much their generality as their mobility.

The instinctive sign is adherent, *the intelligent sign is* mobile.

Now, this mobility of words, that makes them able to pass from one thing to another, has enabled them to be extended from things to ideas. Certainly, language would not have given the faculty of reflecting to an intelligence entirely externalized and incapable of turning homeward. An intelligence which reflects is one that originally had a surplus of energy to spend, over and above practically useful efforts. It is a consciousness that has virtually reconquered itself. But still the virtual has to become actual. Without language, intelligence would probably have remained riveted to the material objects which it was interested in considering. It would have lived in a state of somnambulism, outside itself, hypnotized on its own work. Language has greatly contributed to its liberation. The word, made to pass from one thing to another, is, in fact, by nature transferable and free. It can therefore be extended, not only from one perceived thing to another, but even from a perceived thing to a recollection of that thing, from the precise recollection to a more fleeting image, and finally from an image fleeting, though still pictured, to the picturing of the act by which the image is pictured, that is to say, to the idea. Thus is revealed to the intelligence, hitherto always turned outwards, a whole internal world—the spectacle of its own workings. It required only this opportunity, at length offered by language. It profits by the fact that the word is an external thing, which the intelligence can catch hold of and cling to, and at the same time an immaterial thing, by means of which the intelligence can penetrate even to the inwardness of its own work. Its first business was indeed to make instruments, but this fabrication is possible only by the employment of certain means which are not cut to the exact measure of their object, but go beyond it and thus allow intelligence a supplementary—that is to say disinterested work. From the moment that the intellect, reflecting upon its own doings, perceives itself as a creator of ideas, as a faculty of representation in general, there is no object of which it may not wish to have the idea, even though that object be without direct relation to practical action. That is why we said there are things that intellect alone can seek. Intellect alone, indeed, troubles itself about theory; and its theory would fain embrace everything—not only inanimate matter, over which it has a natural hold, but even life and thought.

By what means, what instruments, in short by what method it will approach these problems, we can easily guess. Originally, it was fashioned to the form of matter. Language itself, which has enabled it to extend its field of operations, is made to designate things, and nought but things: it is only because the word is mobile, because it flies from one thing to another, that the intellect was sure to take it, sooner or later, on the wing, while it was not settled on anything, and apply it to an object which is not a thing and which, concealed till then, awaited the coming of the word to pass from darkness to light. But the word, by covering up this object, again converts it into a thing. So intelligence, even when it no longer operates upon its own object, follows habits it has contracted in that operation: it applies forms that are indeed those of unorganized matter. It is made for this kind of work.

With this kind of work alone is it fully satisfied. And that is what intelligence expresses by saying that thus only it arrives at *distinctness* and *clearness*.

It must, therefore, in order to think itself clearly and distinctly, perceive itself under the form of discontinuity. Concepts, in fact, are outside each other, like objects in space; and they have the same stability as such objects, on which they have been modeled. Taken together, they constitute an "intelligible world," that resembles the world of solids in its essential characters, but whose elements are lighter, more diaphanous, easier for the intellect to deal with than the image of concrete things: they are not, indeed, the perception itself of things, but the representation of the act by which the intellect is fixed on them. They are, therefore, not images, but symbols. Our logic is the complete set of rules that must be followed in using symbols. As these symbols are derived from the consideration of solids, as the rules for combining these symbols hardly do more than express the most general relations among solids, our logic triumphs in that science which takes the solidity of bodies for its object, that is, in geometry. Logic and geometry engender each other, as we shall see a little further on. It is from the extension of a certain natural geometry, suggested by the most general and immediately perceived properties of solids, that natural logic has arisen; then from this natural logic, in its turn, has sprung scientific geometry, which extends further and further the knowledge of the external properties of solids.[1] Geometry and logic are strictly applicable to matter; in it they are at home, and in it they can proceed quite alone. But, outside this do-

[1] We shall return to these points in the next chapter.

main, pure reasoning needs to be supervised by common sense, which is an altogether different thing.

Thus, all the elementary forces of the intellect tend to transform matter into an instrument of action, that is, in the etymological sense of the word, into an *organ*. Life, not content with producing organisms, would fain give them as an appendage inorganic matter itself, converted into an immense organ by the industry of the living being. Such is the initial task it assigns to intelligence. That is why the intellect always behaves as if it were fascinated by the contemplation of inert matter. It is life looking outward, putting itself outside itself, adopting the ways of unorganized nature in principle, in order to direct them in fact. Hence its bewilderment when it turns to the living and is confronted with organization. It does what it can, it resolves the organized into the unorganized, for it cannot, without reversing its natural direction and twisting about on itself, think true continuity, real mobility, reciprocal penetration—in a word, that creative evolution which is life.

Consider continuity. The aspect of life that is accessible to our intellect—as indeed to our senses, of which our intellect is the extension—is that which offers a hold to our action. Now, to modify an object, we have to perceive it as divisible and discontinuous. From the point of view of positive science, an incomparable progress was realized when the organized tissues were resolved into cells. The study of the cell, in its turn, has shown it to be an organism whose complexity seems to grow, the more thoroughly it is examined. The more science advances, the more it sees the number grow of heterogeneous elements which are placed together, outside each other, to make up a living being. Does science thus get any nearer to life? Does it not, on the contrary, find that what is really life in the living seems to recede with every step by which it pushes further the detail of the parts combined? There is indeed already among scientists a tendency to regard the substance of the organism as continuous, and the cell as an artificial entity.[1] But, supposing this view were finally to prevail, it could only lead, on deeper study, to some other mode of analyzing of the living being, and so to a new discontinuity—although less removed, perhaps, from the real continuity of life. The truth is that this continuity cannot be thought by the intellect while it follows its natural movement. It implies at once the multiplicity of elements and the interpenetration of all by all, two conditions that can hardly be reconciled in the field in which our industry, and consequently our intellect, is engaged.

Just as we separate in space, we fix in time. The intellect is not made to think *evolution*, in the proper sense of the word—that is to say, the continuity of a change that is pure mobility. We shall not dwell here on this point, which we propose to study in a special chapter. Suffice it to say that the intellect represents *becoming* as a series of *states*, each of which is homogeneous with itself and consequently does not change. Is our attention called to the internal change of one of these states? At once we decompose it into another series of states which, reunited, will be supposed to make up this internal modification. Each of these new states must be invariable, or else their internal change, if we are forced to notice it, must be resolved again into a fresh series of invariable states, and

[1] We shall return to this point in chapter iii., p. 283.

so on to infinity. Here again, thinking consists in reconstituting, and, naturally, it is with *given* elements, and consequently with *stable* elements, that we reconstitute. So that, though we may do our best to imitate the mobility of becoming by an addition that is ever going on, becoming itself slips through our fingers just when we think we are holding it tight.

Precisely because it is always trying to reconstitute, and to reconstitute with what is given, the intellect lets what is *new* in each moment of a history escape. It does not admit the unforeseeable. It rejects all creation. That definite antecedents bring forth a definite consequent, calculable as a function of them, is what satisfies our intellect. That a definite end calls forth definite means to attain it, is what we also understand. In both cases we have to do with the known which is combined with the known, in short, with the old which is repeated. Our intellect is there at its ease; and, whatever be the object, it will abstract, separate, eliminate, so as to substitute for the object itself, if necessary, an approximate equivalent in which things will happen in this way. But that each instant is a fresh endowment, that the new is ever upspringing, that the form just come into existence (although, *when once produced*, it may be regarded as an effect determined by its causes) could never have been foreseen—because the causes here, unique in their kind, are part of the effect, have come into existence with it, and are determined by it as much as they determine it—all this we can feel within ourselves and also divine, by sympathy, outside ourselves, but we cannot think it, in the strict sense of the word, nor express it in terms of pure understanding. No wonder at that: we must remember what our intellect is meant for. The causality it seeks and finds everywhere expresses the very mechanism of our industry, in which we go on recomposing the same whole with the same parts, repeating the same movements to obtain the same result. The finality it understands best is the finality of our industry, in which we work on a model given in advance, that is to say, old or composed of elements already known. As to invention properly so called, which is, however, the point of departure of industry itself, our intellect does not succeed in grasping it in its *upspringing,* that is to say, in its indivisibility, nor in its *fervor,* that is to say, in its creativeness. Explaining it always consists in resolving it, it the unforeseeable and new, into elements old or known, arranged in a different order. The intellect can no more admit complete novelty than real becoming; that is to say, here again it lets an essential aspect of life escape, as if it were not intended to think such an object.

All our analyses bring us to this conclusion. But it is hardly necessary to go into such long details concerning the mechanism of intellectual working; it is enough to consider the results. We see that the intellect, so skilful in dealing with the inert, is awkward the moment it touches the living. Whether it wants to treat the life of the body or the life of the mind, it proceeds with the rigor, the stiffness and the brutality of an instrument not designed for such use. The history of hygiene or of pedagogy teaches us much in this matter. When we think of the cardinal, urgent and constant need we have to preserve our bodies and to raise our souls, of the special facilities given to each of us, in this field, to experiment continually on ourselves and on others, of the palpable injury by which the wrongness of a medical or pedagogical practice is both made manifest and punished at once, we are amazed at the stupidity and especially at the persistence of errors. We may easily find their origin in the natural obstinacy with which we treat the living like the lifeless and think all reality, however fluid, under the form of the sharply defined solid. We are at ease only in the discontinuous, in the immobile, in the dead. *The intellect is characterized by a natural inability to comprehend life.*

Instinct, on the contrary, is molded on the very form of life. While intelligence treats everything mechanically, instinct proceeds, so to speak, organically. If the consciousness that slumbers in it should awake, if it were wound up into knowledge instead of being wound off into action, if we could ask and it could reply, it would give up to us the most intimate secrets of life. For it only carries out further the work by which life organizes matter —so that we cannot say, as has often been shown, where organization ends and where instinct begins. When the little chick is breaking its shell with a peck of its beak, it is acting by instinct, and yet it does but carry on the movement which has borne it through embryonic life. Inversely, in the course of embryonic life itself (especially when the embryo lives freely in the form of a larva), many of the acts accomplished must be referred to instinct. The most essential of the primary instincts are really, therefore, vital processes. The potential consciousness that accompanies them is generally actualized only at the outset of the act, and leaves the rest of the process to go on by itself. It would only have to expand more widely, and then dive into its own depth completely, to be one with the generative force of life.

When we see in a living body thousands of cells working together to a common end, dividing the task between them, living each for itself at the same time as for the others, preserving itself, feeding itself, reproducing itself, responding to the menace of danger by appropriate defensive reactions, how can we help thinking of so many instincts? And yet these are the natural functions of the cell, the constitutive elements of its vitality. On the other hand, when we see the bees of a hive forming a system so strictly organized that no individual can live apart from the others beyond a certain time, even though furnished with food and shelter, how can we help recognizing that the hive is really, and not metaphorically, a single organism, of which each bee is a cell united to the others by invisible bonds? The instinct that animates the bee is indistinguishable, then, from the force that animates the cell, or is only a prolongation of that force. In extreme cases like this, instinct coincides with the work of organization.

Of course there are degrees of perfection in the same instinct. Between the humble-bee, and the honey-bee, for instance, the distance is great; and we pass from one to the other through a great number of intermediaries, which correspond to so many complications of the social life. But the same diversity is found in the functioning of histological elements belonging to different tissues more or less akin. In both cases there are manifold variations on one and the same theme. The constancy of the theme is manifest, however, and the variations only fit it to the diversity of the circumstances.

Now, in both cases, in the instinct of the animal and in the vital properties of the cell, the same knowledge and the same ignorance are shown. All goes on as if the cell knew, of the other cells, what concerns itself; as if the animal knew, of the other animals, what it can utilize —all else remaining in shade. It seems as if life, as soon as it has become bound up in a species, is cut off from the rest of its own work, save at one or two points that are of vital concern to the species just arisen. Is it not plain that life goes to work here exactly like consciousness, exactly like memory? We trail behind us, unawares, the whole of our past; but our memory pours into the present only the odd recollection or two that in some way complete our present situation. Thus the instinctive knowledge which one species possesses of another on a certain particular point has its root in the very unity of life, which is, to use the expression of an ancient philosopher, a "whole sympathetic to itself." It is impossible to consider some of the special instincts of the animal and of the plant, evidently arisen in extraordinary circumstances, without relating them to those recollections, seemingly forgotten, which spring up suddenly under the pressure of an urgent need.

No doubt many secondary instincts, and also many varieties of primary instinct, admit of a scientific explanation. Yet it is doubtful whether science, with its present methods of explanation, will ever succeed in analyzing instinct completely. The reason is that instinct and intelligence are two divergent developments of one and the same principle, which in the one case remains within itself, in the other steps out of itself and becomes absorbed in the utilization of inert matter. This gradual divergence testifies to a radical incompatibility, and points to the fact that it is impossible for intelligence to re-absorb instinct. That which is instinctive in instinct cannot be expressed in terms of intelligence, nor, consequently, can it be analyzed.

A man born blind, who had lived among others born blind, could not be made to believe in the possibility of perceiving a distant object without first perceiving all the objects in between. Yet vision performs this miracle. In a certain sense the blind man is right, since vision, having its origin in the stimulation of the retina, by the vibrations of the light, is nothing else, in fact, but a retinal touch. Such is indeed the *scientific* explanation, for the function of science is just to express all perceptions in terms of touch. But we have shown elsewhere that the philosophical explanation of perception (if it may still be called an explanation) must be of another kind.[1] Now instinct also is a knowledge at a distance. It has the same relation to intelligence that vision has to touch. Science cannot do otherwise than express it in terms of intelligence; but in so doing it constructs an imitation of instinct rather than penetrates within it.

Anyone can convince himself of this by studying the ingenious theories of evolutionist biology. They may be reduced to two types, which are often intermingled. One type, following the principles of neo-Darwinism, regards instinct as a sum of accidental differences preserved by selection: such and such a useful behavior, naturally adopted by the individual in virtue of an accidental predisposition of the germ, has been transmitted from germ to germ, waiting for chance to add fresh improvements

[1] *Matière et mémoire*, chap. i.

to it by the same method. The other type regards instinct as lapsed intelligence: the action, found useful by the species or by certain of its representatives, is supposed to have engendered a habit, which, by hereditary transmission, has become an instinct. Of these two types of theory, the first has the advantage of being able to bring in hereditary transmission without raising grave objection; for the accidental modification which it places at the origin of the instinct is not supposed to have been acquired by the individual, but to have been inherent in the germ. But, on the other hand, it is absolutely incapable of explaining instincts as sagacious as those of most insects. These instincts surely could not have attained, all at once, their present degree of complexity; they have probably evolved; but, in a hypothesis like that of the neo-Darwinians, the evolution of instinct could have come to pass only by the progressive addition of new pieces which, in some way, by happy accidents, came to fit into the old. Now it is evident that, in most cases, instinct could not have perfected itself by simple accretion: each new piece really requires, if all is not to be spoiled, a complete recasting of the whole. How could mere chance work a recasting of the kind? I agree that an accidental modification of the germ may be passed on hereditarily, and may somehow wait for fresh accidental modifications to come and complicate it. I agree also that natural selection may eliminate all those of the more complicated forms of instinct that are not fit to survive Still, in order that the life of the instinct may evolve, complications fit to survive have to be produced. Now they will be produced only if, in certain cases, the addition of a new element brings about the correlative change of all the old elements. No one will maintain that chance could perform such a miracle: in one form or another we shall appeal to intelligence. We shall suppose that it is by an effort, more or less conscious, that the living being develops a higher instinct. But then we shall have to admit that an acquired habit can become hereditary, and that it does so regularly enough to ensure an evolution. The thing is doubtful, to put it mildly. Even if we could refer the instincts of animals to habits intelligently acquired and hereditarily transmitted, it is not clear how this sort of explanation could be extended to the vegetable world, where effort is never intelligent, even supposing it is sometimes conscious. And yet, when we see with what sureness and precision climbing plants use their tendrils, what marvelously combined manœuvres the orchids perform to procure their fertilization by means of insects,[1] how can we help thinking that these are so many instincts?

This is not saying that the theory of the neo-Darwinians must be altogether rejected, any more than that of the neo-Lamarckians. The first are probably right in holding that evolution takes place from germ to germ rather than from individual to individual; the second are right in saying that at the origin of instinct there is an effort (although it is something quite different, we believe, from an *intelligent* effort). But the former are probably wrong when they make the evolution of instinct an *accidental* evolution, and the latter when they regard the effort from which instinct proceeds as an *individual* effort. The effort by which a species modifies its

[1] See the two works of Darwin, *Climbing Plants* and *The Fertilization of Orchids by Insects*

instinct, and modifies itself as well, must be a much deeper thing, dependent solely neither on circumstances nor on individuals. It is not purely accidental, although accident has a large place in it; and it does not depend solely on the initiative of individuals, although individuals collaborate in it.

Compare the different forms of the same instinct in different species of hymenoptera. The impression derived is not always that of an increasing complexity made of elements that have been added together one after the other. Nor does it suggest the idea of steps up a ladder. Rather do we think, in many cases at least, of the circumference of a circle, from different points of which these different varieties have started, all facing the same center, all making an effort in that direction, but each approaching it only to the extent of its means, and to the extent also to which this central point has been illumined for it. In other words, instinct is everywhere complete, but it is more or less simplified, and, above all, simplified *differently*. On the other hand, in cases where we do get the impression of an ascending scale, as if one and the same instinct had gone on complicating itself more and more in one direction and along a straight line, the species which are thus arranged by their instincts into a linear series are by no means always akin. Thus, the comparative study, in recent years, of the social instinct in the different apidae proves that the instinct of the meliponines is intermediary in complexity between the still rudimentary tendency of the humble bees and the consummate science of the true bees; yet there can be no kinship between the bees and the meliponines.[1]

[1] Buttel-Reepen, "Die phylogenetische Entstehung des Bienenstaates" (*Biol. Centralblatt*, xxiii. 1903), p. 108 in particular.

Most likely, the degree of complexity of these different societies has nothing to do with any greater or smaller number of added elements. We seem rather to be before a *musical theme*, which had first been transposed, the theme as a whole, into a certain number of tones, and on which, still the whole theme, different variations had been played, some very simple, others very skilful. As to the original theme, it is everywhere and nowhere. It is in vain that we try to express it in terms of any idea: it must have been, originally, *felt* rather than *thought*. We get the same impression before the paralyzing instinct of certain wasps. We know that the different species of hymenoptera that have this paralyzing instinct lay their eggs in spiders, beetles or caterpillars, which, having first been subjected by the wasp to a skilful surgical operation, will go on living motionless a certain number of days, and thus provide the larvae with fresh meat. In the sting which they give to the nerve-centers of their victim, in order to destroy its power of moving without killing it, these different species of hymenoptera take into account, so to speak, the different species of prey they respectively attack. The Scolia, which attacks a larva of the rose-beetle, stings it in one point only, but in this point the motor ganglia are concentrated, and those ganglia alone: the stinging of other ganglia might cause death and putrefaction, which it must avoid.[1] The yellow-winged Sphex, which has chosen the cricket for its victim, knows that the cricket has three nerve-centers which serve its three pairs of legs—or at least it acts as if it knew this. It stings the insect first under the neck, then behind the prothorax, and then where the thorax

[1] Fabre, *Souvenirs entomologiques*, 3e série, Paris, 1890, pp. 1-69.

joins the abdomen.[1] The Ammophila Hirsuta gives nine successive strokes of its sting upon nine nerve-centers of its caterpillar, and then seizes the head and squeezes it in its mandibles, enough to cause paralysis without death.[2] The general theme is "the necessity of paralyzing without killing"; the variations are subordinated to the structure of the victim on which they are played. No doubt the operation is not always perfect. It has recently been shown that the Ammophila sometimes kills the caterpillar instead of paralyzing it, that sometimes also it paralyzes it incompletely.[3] But, because instinct is, like intelligence, fallible, because it also shows individual deviations, it does not at all follow that the instinct of the Ammophila has been acquired, as has been claimed, by tentative intelligent experiments. Even supposing that the Ammophila has come in course of time to recognize, one after another, by tentative experiment, the points of its victim which must be stung to render it motionless, and also the special treatment that must be inflicted on the head to bring about paralysis without death, how can we imagine that elements so special of a knowledge so precise have been regularly transmitted, one by one, by heredity? If, in all our present experience, there were a single indisputable example of a transmission of this kind, the inheritance of acquired characters would be questioned by no one. As a matter of fact, the hereditary transmission of a contracted habit is effected in an irregular and far from precise manner, supposing it is ever really effected at all.

[1] Fabre, *Souvenirs entomologiques*, 1re série, 3e édition, Paris, 1894, pp. 93 ff.

[2] Fabre, *Nouveaux souvenirs entomologiques,* Paris, 1882, pp. 14 ff.

[3] Peckham, *Wasps, Solitary and Social*, Westminster, 1905, pp. 28 ff.

But the whole difficulty comes from our desire to express the knowledge of the hymenoptera in terms of intelligence. It is this that compels us to compare the Ammophila with the entomologist, who knows the caterpillar as he knows everything else—from the outside, and without having on his part a special or vital interest. The Ammophila, we imagine, must learn, one by one, like the entomologist, the positions of the nerve-centers of the caterpillar—must acquire at least the practical knowledge of these positions by trying the effects of its sting. But there is no need for such a view if we suppose a *sympathy* (in the etymological sense of the word) between the Ammophila and its victim, which teaches it from within, so to say, concerning the vulnerability of the caterpillar. This feeling of vulnerability might owe nothing to outward perception, but result from the mere presence together of the Ammophila and the caterpillar, considered no longer as two organisms, but as two activities. It would express, in a concrete form, the *relation* of the one to the other. Certainly, a scientific theory cannot appeal to considerations of this kind. It must not put action before organization, sympathy before perception and knowledge. But, once more, either philosophy has nothing to see here, or its rôle begins where that of science ends.

Whether it makes instinct a "compound reflex," or a habit formed intelligently that has become automatism, or a sum of small accidental advantages accumulated and fixed by selection, in every case science claims to resolve instinct completely either into *intelligent* actions, or into mechanisms built up piece by piece like those combined by our *intelligence*. I agree indeed that science is here within its function. It gives us, in default of a real analysis of the object, a translation of this object in terms of intelligence. But is it not plain that science itself invites philosophy to consider things in another way? If our biology was still that of Aristotle, if it regarded the series of living beings as unilinear, if it showed us the whole of life evolving toward intelligence and passing, to that end, through sensibility and instinct, we should be right, we, the intelligent beings, in turning back toward the earlier and consequently inferior manifestations of life and in claiming to fit them, without deforming them, into the molds of our understanding. But one of the clearest results of biology has been to show that evolution has taken place along divergent lines. It is at the extremity of two of these lines—the two principal—that we find intelligence and instinct in forms almost pure. Why, then, should instinct be resolvable into intelligent elements? Why, even, into terms entirely intelligible? Is it not obvious that to think here of the intelligent, or of the absolutely intelligible, is to go back to the Aristotelian theory of nature? No doubt it is better to go back to that than to stop short before instinct as before an unfathomable mystery. But, though instinct is not within the domain of intelligence, it is not situated beyond the limits of mind. In the phenomena of feeling, in unreflecting sympathy and antipathy, we experience in ourselves—though under a much vaguer form, and one too much penetrated with intelligence—something of what must happen in the consciousness of an insect acting by instinct. Evolution does but sunder, in order to develop them to the end, elements which, at their origin, interpenetrated each other. More precisely, intelligence is, before anything else, the faculty of relating one point of space to another, one material object to another; it applies to all things, but remains outside them; and of a deep cause it perceives only the effects spread out side by side. Whatever be the force that is at work in the genesis of the nervous system of the caterpillar, to our eyes and our intelligence it is only a juxtaposition of nerves and nervous centers. It is true that we thus get the whole outer effect of it. The Ammophila, no doubt, discerns but a very little of that force, just what concerns itself; but at least it discerns it from within, quite otherwise than by a process of knowledge—by an intuition (*lived* rather than *represented*), which is probably like what we call divining sympathy.

A very significant fact is the swing to and fro of scientific theories of instinct, from regarding it as intelligent to regarding it as simply intelligible, or, shall I say, between likening it to an intelligence "lapsed" and reducing it to a pure mechanism.[1] Each of these systems of explanation triumphs in its criticism of the other, the first when it shows us that instinct cannot be a mere reflex, the other when it declares that instinct is something different from intelligence, even fallen into unconsciousness. What can this mean but that they are two symbolisms, equally acceptable in certain respects, and, in other respects, equally inadequate to their object? The concrete explanation, no longer scientific, but metaphysical, must be sought along quite another path, not in the direction of intelligence, but in that of "sympathy."

[1] See, in particular, among recent works, Bethe, "Dürfen wir den Ameisen und Bienen psychische Qualitäten zuschreiben?" (*Arch. f. d. ges. Physiologie*, 1898), and Forel, "Un Aperçu de psychologie comparée" (*Année psychologique*, 1895).

Instinct is sympathy. If this sympathy could extend its object and also reflect upon itself, it would give us the key to vital operations—just as intelligence, developed and disciplined, guides us into matter. For—we cannot too often repeat it—intelligence and instinct are turned in opposite directions, the former toward inert matter, the latter toward life. Intelligence, by means of science, which is its work, will deliver up to us more and more completely the secret of physical operations; of life it brings us, and moreover only claims to bring us, a translation in terms of inertia. It goes all round life, taking from outside the greatest possible number of views of it, drawing it into itself instead of entering into it. But it is to the very inwardness of life that *intuition* leads us—by intuition I mean instinct that has become disinterested, self-conscious, capable of reflecting upon its object and of enlarging it indefinitely.

That an effort of this kind is not impossible, is proved by the existence in man of an aesthetic faculty along with normal perception. Our eye perceives the features of the living being, merely as assembled, not as mutually organized. The intention of life, the simple movement that runs through the lines, that binds them together and gives them significance, escapes it. This intention is just what the artist tries to regain, in placing himself back within the object by a kind of sympathy, in breaking down, by an effort of intuition, the barrier that space puts up between him and his model. It is true that this aesthetic intuition, like external perception, only attains the individual. But we can conceive an inquiry turned in the same direction as art, which would take life *in general* for its object, just as physical science, in following to the end the direction pointed out by external perception, prolongs the individual facts into general laws. No doubt this philosophy will never obtain a knowledge of its object comparable to that which science has of its own. Intelligence remains the luminous nucleus around which instinct, even enlarged and purified into intuition, forms only a vague nebulosity. But, in default of knowledge properly so called, reserved to pure intelligence, intuition may enable us to grasp what it is that intelligence fails to give us, and indicate the means of supplementing it. On the one hand, it will utilize the mechanism of intelligence itself to show how intellectual molds cease to be strictly applicable; and on the other hand, by its own work, it will suggest to us the vague feeling, if nothing more, of what must take the place of intellectual molds. Thus, intuition may bring the intellect to recognize that life does not quite go into the category of the many nor yet into that of the one; that neither mechanical causality nor finality can give a sufficient interpretation of the vital process. Then, by the sympathetic communication which it establishes between us and the rest of the living, by the expansion of our consciousness which it brings about, it introduces us into life's own domain, which is reciprocal interpenetration, endlessly continued creation. But, though it thereby transcends intelligence, it is from intelligence that has come the push that has made it rise to the point it has reached. Without intelligence, it would have remained in the form of instinct, riveted to the special object of its practical interest, and turned outward by it into movements of locomotion.

How theory of knowledge must take account of these two faculties, intellect and intuition, and how also, for want of establishing a sufficiently clear distinction between them, it becomes involved in inextricable difficulties, creating phantoms of ideas to which there cling phantoms of problems, we shall endeavor to show a little further on. We shall see that the problem of knowledge, from this point of view, is one with the metaphysical problem, and that both one and the other depend upon experience. On the one hand, indeed, if intelligence is charged with matter and instinct with life, we must squeeze them both in order to get the double essence from them; metaphysics is therefore dependent upon theory of knowledge. But, on the other hand, if consciousness has thus split up into intuition and intelligence, it is because of the need it had to apply itself to matter at the same time as it had to follow the stream of life. The double form of consciousness is then due to the double form of the real, and theory of knowledge must be dependent upon metaphysics. In fact, each of these two lines of thought leads to the other; they form a circle, and there can be no other center to the circle but the empirical study of evolution. It is only in seeing consciousness run through matter, lose itself there and find itself there again, divide and reconstitute itself, that we shall form an idea of the mutual opposition of the two terms, as also, perhaps, of their common origin. But, on the other hand, by dwelling on this opposition of the two elements and on this identity of origin, perhaps we shall bring out more clearly the meaning of evolution itself.

Such will be the aim of our next chapter. But the facts that we have just noticed must have already suggested to us the idea that life is connected either with consciousness or with something that resembles it.

Throughout the whole extent of the animal kingdom, we have said, consciousness seems proportionate to the living being's power of choice. It lights up the zone of potentialities that surrounds the act. It fills the interval between what is done and what might be done. Looked at from without, we may regard it as a simple aid to action, a light that action kindles, a momentary spark flying up from the friction of real action against possible actions. But we must also point out that things would go on in just the same way if consciousness, instead of being the effect, were the cause. We might suppose that consciousness, even in the most rudimentary animal, covers by right an enormous field, but is compressed in fact in a kind of vise: each advance of the nervous centers, by giving the organism a choice between a larger number of actions, calls forth the potentialities that are capable of surrounding the real, thus opening the vise wider and allowing consciousness to pass more freely. In this second hypothesis, as in the first, consciousness is still the instrument of action; but it is even more true to say that action is the instrument of consciousness; for the complicating of action with action, and the opposing of action to action, are for the imprisoned consciousness the only possible means to set itself free. How, then, shall we choose between the two hypotheses? If the first is true, consciousness must express exactly, at each instant, the state of the brain; there is strict parallelism (so far as intelligible) between the psychical and the cerebral state. On the second hypothesis, on the contrary, there is indeed solidarity and interdependence between the brain and consciousness, but not parallelism: the more complicated the brain becomes, thus giving the organism greater choice of possible actions, the more does consciousness outrun its physical concomitant. Thus, the recollection of the same spectacle probably modifies in the same way a dog's brain and a man's brain, if the perception has been the same; yet the recollection must be very different in the man's consciousness from what it is in the dog's. In the dog, the recollection remains the captive of perception; it is brought back to consciousness only when an analogous perception recalls it by reproducing the same spectacle, and then it is manifested by the recognition, *acted* rather than *thought*, of the present perception much more than by an actual reappearance of the recollection itself. Man, on the contrary, is capable of calling up the recollection at will, at any moment, independently of the present perception. He is not limited to *playing* his past life again; he *represents* and *dreams* it. The local modification of the brain to which the recollection is attached being the same in each case, the psychological difference between the two recollections cannot have its ground in a particular difference of detail between the two cerebral mechanisms, but in the difference between the two brains taken each as a whole. The more complex of the two, in putting a greater number of mechanisms in opposition to one another, has enabled consciousness to disengage itself from the restraint of one and all and to reach independence. That things do happen in this way, that the second of the two hypotheses is that which must be chosen, is what we have tried to prove, in a former work, by the study of facts that best bring into relief the relation of the conscious state to the cerebral state, the facts of normal and pathological recognition, in particular the forms of aphasia.[1] But it could have been proved by pure reasoning, before even it was evidenced by facts. We have shown on what self-contradictory postulate, on what confusion of two mutually incompatible symbolisms, the hypothesis of equivalence between the cerebral state and the psychic state rests.[2]

The evolution of life, looked at from this point, receives a clearer meaning, although it cannot be subsumed under any actual *idea*. It is as if a broad current of consciousness had penetrated matter, loaded, as all consciousness is, with an enormous multiplicity of interwoven potentialities. It has carried matter along to organization, but its movement has been at once infinitely retarded and infinitely divided. On the one hand, indeed, consciousness has had to fall asleep, like the chrysalis in the envelope in which it is preparing for itself wings; and, on the other hand, the manifold tendencies it contained have been distributed among divergent series of organisms which, moreover, express these tendencies outwardly in movements rather than internally in representations. In the course of this evolution, while some beings have fallen more and more asleep, others have more and more completely awakened, and the torpor of some has served the activity of others. But the waking could be effected in two different ways. Life, that is to say consciousness launched into matter, fixed its attention either on its own movement or on the matter it was passing through; and it has thus been turned either in the direction of intuition or in that of intellect. Intuition, at

[1] *Matière et mémoire,* chaps. ii. and iii.

[2] "Le Paralogisme psycho-physiologique" (*Revue de métaphysique,* Nov. 1904).

first sight, seems far preferable to intellect, since in it life and consciousness remain within themselves. But a glance at the evolution of living beings shows us that intuition could not go very far. On the side of intuition, consciousness found itself so restricted by its envelope that intuition had to shrink into instinct, that is, to embrace only the very small portion of life that interested it; and this it embraces only in the dark, touching it while hardly seeing it. On this side, the horizon was soon shut out. On the contrary, consciousness, in shaping itself into intelligence, that is to say in concentrating itself at first on matter, seems to externalize itself in relation to itself; but, just because it adapts itself thereby to objects from without, it succeeds in moving among them and in evading the barriers they oppose to it, thus opening to itself an unlimited field. Once freed, moreover, it can turn inwards on itself, and awaken the potentialities of intuition which still slumber within it.

From this point of view, not only does consciousness appear as the motive principle of evolution, but also, among conscious beings themselves, man comes to occupy a privileged place. Between him and the animals the difference is no longer one of degree, but of kind. We shall show how this conclusion is arrived at in our next chapter. Let us now show how the preceding analyses suggest it.

A noteworthy fact is the extraordinary disproportion between the consequences of an invention and the invention itself. We have said that intelligence is modeled on matter and that it aims in the first place at fabrication. But does it fabricate in order to fabricate or does it not pursue involuntarily, and even unconsciously, something entirely different? Fabricating consists in shaping matter, in making it supple and in bending it, in converting it into an instrument in order to become master of it. It is this *mastery* that profits humanity, much more even than the material result of the invention itself. Though we derive an immediate advantage from the thing made, as an intelligent animal might do, and though this advantage be all the inventor sought, it is a slight matter compared with the new ideas and new feelings that the invention may give rise to in every direction, as if the essential part of the effect were to raise us above ourselves and enlarge our horizon. Between the effect and the cause the disproportion is so great that it is difficult to regard the cause as *producer* of its effect. It releases it, whilst settling, indeed, its direction. Everything happens as though the grip of intelligence on matter were, in its main intention, to *let something pass* that matter is holding back.

The same impression arises when we compare the brain of man with that of the animals. The difference at first appears to be only a difference of size and complexity. But, judging by function, there must be something else besides. In the animal, the motor mechanisms that the brain succeeds in setting up, or, in other words, the habits contracted voluntarily, have no other object nor effect than the accomplishment of the movements marked out in these habits, stored in these mechanisms. But, in man, the motor habit may have a second result, out of proportion to the first: it can hold other motor habits in check, and thereby, in overcoming automatism, set consciousness free. We know what vast regions in the human brain language occupies. The cerebral mechanisms that correspond to the words have this in particular, that they can be made to grapple with other mechanisms, those, for instance, that correspond to the things themselves, or even be made to grapple with one another. Meanwhile consciousness, which would have been dragged down and drowned in the accomplishment of the act, is restored and set free.[1]

The difference must therefore be more radical than a superficial examination would lead us to suppose. It is the difference between a mechanism which engages the attention and a mechanism from which it can be diverted. The primitive steam engine, as Newcomen conceived it, required the presence of a person exclusively employed to turn on and off the taps, either to let the steam into the cylinder or to throw the cold spray into it in order to condense the steam. It is said that a boy employed on this work, and very tired of having to do it, got the idea of tying the handles of the taps, with cords, to the beam of the engine. Then the machine opened and closed the taps itself; it worked all alone. Now, if an observer had compared the structure of this second machine with that of the first without taking into account the two boys left to watch over them, he would have found only a slight difference of complexity. That is, indeed, all we can perceive when we look only at the machines. But if we cast a glance at the two boys, we shall see that whilst one is wholly taken up by the watching,

[1] A geologist whom we have already had occasion to cite, N. S. Shaler, well says that "when we come to man, it seems as if we find the ancient subjection of mind to body abolished, and the intellectual parts develop with an extraordinary rapidity, the structure of the body remaining identical in essentials" (Shaler, *The Interpretation of Nature*, Boston, 1899, p. 187).

the other is free to go and play as he chooses, and that, from this point of view, the difference between the two machines is radical, the first holding the attention captive, the second setting it at liberty. A difference of the same kind, we think, would be found between the brain of an animal and the human brain.

If, now, we should wish to express this in terms of finality, we should have to say that consciousness, after having been obliged, in order to set itself free, to divide organization into two complementary parts, vegetables on one hand and animals on the other, has sought an issue in the double direction of instinct and of intelligence. It has not found it with instinct, and it has not obtained it on the side of intelligence except by a sudden leap from the animal to man. So that, in the last analysis, man might be considered the reason for the existence of the entire organization of life on our planet. But this would be only a manner of speaking. There is, in reality, only a current of existence and the opposing current: thence proceeds the whole evolution of life. We must now grasp more closely the opposition of these two currents. Perhaps we shall thus discover for them a common source. By this we shall also, no doubt, penetrate the most obscure regions of metaphysics. However, as the two directions we have to follow are clearly marked, in intelligence on the one hand, in instinct and intuition on the other, we are not afraid of straying. A survey of the evolution of life suggests to us a certain conception of knowledge, and also a certain metaphysics, which imply each other. Once made clear, this metaphysics and this critique may throw some light, in their turn, on evolution as a whole.

CHAPTER III

ON THE MEANING OF LIFE—THE ORDER OF NATURE AND THE FORM OF INTELLIGENCE

In the course of our first chapter we traced a line of demarcation between the inorganic and the organized, but we pointed out that the division of unorganized matter into separate bodies is relative to our senses and to our intellect, and that matter, looked at as an undivided whole, must be a flux rather than a thing. In this we were preparing the way for a reconciliation between the inert and the living.

On the other side, we have shown in our second chapter that the same opposition is found again between instinct and intelligence, the one turned to certain determinations of life, the other molded on the configuration of matter. But instinct and intelligence, we have also said, stand out from the same background, which, for want of a better name, we may call consciousness in general, and which must be coextensive with universal life. In this way, we have disclosed the possibility of showing the genesis of intelligence in setting out from general consciousness, which embraces it.

We are now, then, to attempt a genesis of intellect at the same time as a genesis of material bodies—two enterprises that are evidently correlative, if it be true that the main lines of our intellect mark out the general form of our action on matter, and that the detail of matter is ruled by the requirements of our action. Intellectuality and materiality have been constituted, in detail, by reciprocal adaptation. Both are derived from a wider and higher form of existence. It is there that we must replace them, in order to see them issue forth.

Such an attempt may appear, at first, more daring than the boldest speculations of metaphysicians. It claims to go further than psychology, further than cosmology, further than traditional metaphysics; for psychology, cosmology and metaphysics take intelligence, in all that is essential to it, as given, instead of, as we now propose, engendering it in its form and in its matter. The enterprise is in reality much more modest, as we are going to show. But let us first say how it differs from others.

To begin with psychology, we are not to believe that it *engenders* intelligence when it follows the progressive development of it through the animal series. Comparative psychology teaches us that the more an animal is intelligent, the more it tends to reflect on the actions by which it makes use of things, and thus to approximate to man. But its actions have already by themselves adopted the principal lines of human action; they have made out the same general directions in the material world as we have; they depend upon the same objects bound together by the same relations; so that animal intelligence, although it does not form concepts properly so called, already moves in a conceptual atmosphere. Absorbed at every instant by the actions it performs and the attitudes it must adopt, drawn outward by them and so externalized in relation to itself, it no doubt plays rather than thinks its ideas; this play none the less already corresponds, in the main, to the general plan of human intelligence.[1] To explain the intelligence of man by that of the animal consists then simply in following the development of an embryo of humanity into complete humanity. We show how a certain direction has been followed further and further by beings more and more intelligent. But the moment we admit the direction, intelligence is given.

In a cosmogony like that of Spencer, intelligence is taken for granted, as matter also at the same time. We are shown matter obeying laws, objects connected with objects and facts with facts by constant relations, consciousness receiving the imprint of these relations and laws, and thus adopting the general configuration of nature and shaping itself into intellect. But how can we fail to see that intelligence is supposed when we admit objects and facts? *A priori* and apart from any hypothesis on the nature of the matter, it is evident that the materiality of a body does not stop at the point at which we touch it: a body is present wherever its influence is felt; its attractive force, to speak only of that, is exerted on the sun, on the planets, perhaps on the entire universe. The more physics advances, the more it effaces the individuality of bodies and even of the particles into which the scientific imagination began by decomposing them: bodies and corpuscles tend to dissolve into a universal interaction. Our perceptions give us the plan of our eventual action on things much more than that of things themselves. The outlines we find in objects simply

[1] We have developed this point in *Matière et mémoire*, chaps. ii. and iii., notably pp. 78-80 and 169-186.

mark what we can attain and modify in them. The lines we see traced through matter are just the paths on which we are called to move. Outlines and paths have declared themselves in the measure and proportion that consciousness has prepared for action on unorganized matter—that is to say, in the measure and proportion that intelligence has been formed. It is doubtful whether animals built on a different plan—a mollusk or an insect, for instance—cut matter up along the same articulations. It is not indeed necessary that they should separate it into bodies at all. In order to follow the indications of instinct, there is no need to perceive *objects*, it is enough to distinguish *properties*. Intelligence, on the contrary, even in its humblest form, already aims at getting matter to act on matter. If on one side matter lends itself to a division into active and passive bodies, or more simply into coexistent and distinct fragments, it is from this side that intelligence will regard it; and the more it busies itself with dividing, the more it will spread out in space, in the form of extension adjoining extension, a matter that undoubtedly itself has a tendency to spatiality, but whose parts are yet in a state of reciprocal implication and interpenetration. Thus the same movement by which the mind is brought to form itself into intellect, that is to say, into distinct concepts, brings matter to break itself up into objects excluding one another. *The more consciousness is intellectualized, the more is matter spatialized.* So that the evolutionist philosophy, when it imagines in space a matter cut up on the very lines that our action will follow, has given itself in advance, ready made, the intelligence of which it claims to show the genesis.

Metaphysics applies itself to a work of the same kind, though subtler and more self-conscious, when it deduces *a priori* the categories of thought. It compresses intellect, reduces it to its quintessence, holds it tight in a principle so simple that it can be thought empty: from this principle we then draw out what we have virtually put into it. In this way we may no doubt show the coherence of intelligence, define intellect, give its formula, but we do not trace its genesis. An enterprise like that of Fichte, although more philosophical than that of Spencer, in that it pays more respect to the true order of things, hardly leads us any further. Fichte takes thought in a concentrated state, and expands it into reality; Spencer starts from external reality, and condenses it into intellect. But, in the one case as in the other, the intellect must be taken at the beginning as given—either condensed or expanded, grasped in itself by a direct vision or perceived by reflection in nature, as in a mirror.

The agreement of most philosophers on this point comes from the fact that they are at one in affirming the unity of nature, and in representing this unity under an abstract and geometrical form. Between the organized and the unorganized they do not see and they will not see the cleft. Some start from the inorganic, and, by compounding it with itself, claim to form the living; others place life first, and proceed toward matter by a skilfully managed *decrescendo*; but, for both, there are only differences of *degree* in nature—degrees of complexity in the first hypothesis, of intensity in the second. Once this principle is admitted, intelligence becomes as vast as reality; for it is unquestionable that whatever is geometrical in things is entirely accessible to human intelligence, and if the continuity between geometry and the rest is perfect, all the rest must indeed be equally intelligible, equally intelligent. Such is the postulate of most systems. Anyone can easily be convinced of this by comparing doctrines that seem to have no common point, no common measure, those of Fichte and Spencer for instance, two names that we happen to have just brought together.

At the root of these speculations, then, there are the two convictions correlative and complementary, that nature is one and that the function of intellect is to embrace it in its entirety. The faculty of knowing being supposed coextensive with the whole of experience, there can no longer be any question of engendering it. It is already given, and we merely have to use it, as we use our sight to take in the horizon. It is true that opinions differ as to the value of the result. For some, it is reality itself that the intellect embraces; for others, it is only a phantom. But, phantom or reality, what intelligence grasps is thought to be all that can be attained.

Hence the exaggerated confidence of philosophy in the powers of the individual mind. Whether it is dogmatic or critical, whether it admits the relativity of our knowledge or claims to be established within the absolute, a philosophy is generally the work of a philosopher, a single and unitary vision of the whole. It is to be taken or left.

More modest, and also alone capable of being completed and perfected, is the philosophy we advocate. Human intelligence, as we represent it, is not at all what Plato taught in the allegory of the cave. Its function is not to look at passing shadows nor yet to turn itself round and contemplate the glaring sun. It has something else to do. Harnessed, like yoked oxen, to a heavy task, we feel the play of our muscles and joints, the weight of the plow and the resistance of the soil. To act and to know that we are acting, to come into touch with reality and even to live it, but only in the measure in which it concerns the work that is being accomplished and the furrow that is being plowed, such is the function of human intelligence. Yet a beneficent fluid bathes us, whence we draw the very force to labor and to live. From this ocean of life, in which we are immersed, we are continually drawing something, and we feel that our being, or at least the intellect that guides it, has been formed therein by a kind of local concentration. Philosophy can only be an effort to dissolve again into the Whole. Intelligence, reabsorbed into its principle, may thus live back again its own genesis. But the enterprise cannot be achieved in one stroke; it is necessarily collective and progressive. It consists in an interchange of impressions which, correcting and adding to each other, will end by expanding the humanity in us and making us even transcend it.

But this method has against it the most inveterate habits of the mind. It at once suggests the idea of a vicious circle. In vain, we shall be told, you claim to go beyond intelligence: how can you do that except by intelligence? All that is clear in your consciousness is intelligence. You are inside your own thought; you cannot get out of it. Say, if you like, that the intellect is capable of progress, that it will see more and more clearly into a greater and greater number of things; but do not speak of engendering it, for it is with your intellect itself that you would have to do the work.

The objection presents itself naturally to the mind. But the same reasoning would prove also the impossibility of acquiring any new habit. It is of the essence of reasoning to shut us up in the circle of the given. But action breaks the circle. If we had never seen a man swim, we might say that swimming is an impossible thing, inasmuch as, to learn to swim, we must begin by holding ourselves up in the water and, consequently, already know how to swim. Reasoning, in fact, always nails us down to the solid ground. But if, quite simply, I throw myself into the water without fear, I may keep myself up well enough at first by merely struggling, and gradually adapt myself to the new environment: I shall thus have learnt to swim. So, in theory, there is a kind of absurdity in trying to know otherwise than by intelligence; but if the risk be frankly accepted, action will perhaps cut the knot that reasoning has tied and will not unloose.

Besides, the risk will appear to grow less, the more our point of view is adopted. We have shown that intellect has detached itself from a vastly wider reality, but that there has never been a clean cut between the two; all around conceptual thought there remains an indistinct fringe which recalls its origin. And further we compared the intellect to a solid nucleus formed by means of condensation. This nucleus does not differ radically from the fluid surrounding it. It can only be reabsorbed in it because it is made of the same substance. He who throws himself into the water, having known only the resistance of the solid earth, will immediately be drowned if he does not struggle against the fluidity of the new environment: he must perforce still cling to that solidity, so to speak, which even water presents. Only on this condition can he get used to the fluid's fluidity. So of our thought, when it has decided to make the leap.

But leap it must, that is, leave its own environment. Reason, reasoning on its powers, will never succeed in extending them, though the extension would not appear at all unreasonable once it were accomplished. Thousands and thousands of variations on the theme of walking will never yield a rule for swimming: come, enter the water, and when you know how to swim, you will understand how the mechanism of swimming is connected with that of walking. Swimming is an extension of walking, but walking would never have pushed you on to swimming. So you may speculate as intelligently as you will on the mechanism of intelligence; you will never, by this method, succeed in going beyond it. You may get something more complex, but not something higher nor even something different. You must take things by storm: you must thrust intelligence outside itself by an act of will.

So the vicious circle is only apparent. It is, on the contrary, real, we think, in every other method of philosophy. This we must try to show in a few words, if only to prove that philosophy cannot and must not accept the relation established by pure intellectualism between the theory of knowledge and the theory of the known, between metaphysics and science.

At first sight, it may seem prudent to leave the consideration of facts to positive science, to let physics and chemistry busy themselves with matter, the biological and psychological sciences with life. The task of the philosopher is then clearly defined. He takes facts and laws from the scientists' hand; and whether he tries to go beyond them in order to reach their deeper causes, or whether he thinks it impossible to go further and even proves it by the analysis of scientific knowledge, in both cases he has for the facts and relations, handed over by science, the sort of respect that is due to a final verdict. To this knowledge he adds a critique of the faculty of knowing, and also, if he thinks proper, a metaphysic; but the *matter* of knowledge he regards as the affair of science and not of philosophy.

But how does he fail to see that the real result of this so-called division of labor is to mix up everything and confuse everything? The metaphysic or the critique that the philosopher has reserved for himself he has to receive, ready-made, from positive science, it being already contained in the descriptions and analyses, the whole care of which he left to the scientists. For not having wished to intervene, at the beginning, in questions of fact, he finds himself reduced, in questions of principle, to formulating purely and simply in more precise terms the unconscious and consequently inconsistent metaphysic and critique which the very attitude of science to reality marks out. Let us not be deceived by an apparent analogy between natural things and human things. Here we are not in the judiciary domain, where the description of fact and the judgment on the fact are two distinct things, distinct for the very simple reason that above the fact, and independent of it, there is a law promulgated by a legislator. Here the laws are internal to the facts and relative to the lines that have been followed in cutting the real into distinct facts. We cannot describe the outward appearance of the object without prejudging its inner nature and its organization. Form is no longer entirely isolable from matter, and he who has begun by reserving to philosophy questions of principle, and who has thereby tried to put philosophy above the sciences, as a "court of cassation" is above the courts of assizes and of appeal, will gradually come to make no more of philosophy than a registration court, charged at most with wording more precisely the sentences that are brought to it, pronounced and irrevocable.

Positive science is, in fact, a work of pure intellect. Now, whether our conception of the intellect be accepted or rejected, there is one point on which everybody will agree with us, and that is that the intellect is at home in the presence of unorganized matter. This matter it makes use of more and more by mechanical inventions, and mechanical inventions become the easier to it the more it thinks matter as mechanism. The intellect bears within itself, in the form of natural logic, a latent geometrism that is set free in the measure and proportion that the intellect penetrates into the inner nature of inert matter. Intelligence is in tune with this matter, and that is why the physics and metaphysics of inert matter are so near each other. Now, when the intellect undertakes the study of life, it necessarily treats the living like the inert, applying the same forms to this new object, carrying over into this new field the same habits that have succeeded so well in the old; and it is right to do so, for only on such terms does the living offer to our action the same hold as inert matter. But the truth we thus arrive at becomes altogether relative to our faculty of action. It is no more than a *symbolic* verity. It cannot have the same value as the physical verity, being only an extension of physics to an object which we are *a priori* agreed to look at only in its external aspect. The duty of philosophy should be to intervene here actively, to examine the living without any reservation as to practical utility, by freeing itself from forms and habits that are strictly intellectual. Its own special object is to speculate, that is to say, to see; its attitude toward the living should not be that of science, which aims only at action, and which, being able to act only by means of inert matter, presents to itself the rest of reality in this single respect. What must the result be, if it leave biological and psychological facts to positive science alone, as it has left, and rightly left, physical facts? It will accept *a priori* a mechanistic conception of all nature, a conception unreflected and even unconscious, the outcome of the material need. It will *a priori* accept the doctrine of the simple unity of knowledge and of the abstract unity of nature.

The moment it does so, its fate is sealed. The philosopher has no longer any choice save between a metaphysical dogmatism and a metaphysical skepticism, both of which rest, at bottom, on the same postulate, and neither of which adds anything to positive science. He may hypostasize the unity of nature, or, what comes to the same thing, the unity of science, in a being who is nothing since he does nothing, an ineffectual God who simply sums up in himself all the given; or in an eternal Matter from whose womb have been poured out the properties of things and the laws of nature; or, again, in a pure Form which endeavors to seize an unseizable multiplicity, and which is, as we will, the form of nature or the form of thought. All these philosophies tell us, in their different languages, that science is right to treat the liv ing as the inert, and that there is no difference of value, no distinction to be made between the results which intellect arrives at in applying its categories, whether it rests on inert matter or attacks life.

In many cases, however, we feel the frame cracking. But as we did not begin by distinguishing between the inert and the living, the one adapted in advance to the frame in which we insert it, the other incapable of being held in the frame otherwise than by a convention which eliminates from it all that is essential, we find ourselves, in the end, reduced to regarding everything the frame contains with equal suspicion. To a metaphysical dogmatism, which has erected into an absolute the factitious unity of science, there succeeds a skepticism or a relativism that universalizes and extends to all the results of science the artificial character of some among them. So philosophy swings to and fro between the doctrine that regards absolute reality as unknowable and that which, in the idea it gives us of this reality, says nothing more than science has said. For having wished to prevent all conflict between science and philosophy, we have sacrificed philosophy without any appreciable gain to science. And for having tried to avoid the seeming vicious circle which consists in using the intellect to transcend the intellect, we find ourselves turning in a real circle, that which consists in laboriously rediscovering by metaphysics a unity that we began by positing *a priori*, a unity that we admitted blindly and unconsciously by the very act of abandoning the whole of experience to science and the whole of reality to the pure understanding.

Let us begin, on the contrary, by tracing a line of demarcation between the inert and the living. We shall find that the inert enters naturally into the frames of the intellect, but that the living is adapted to these frames only artificially, so that we must adopt a special attitude toward it and examine it with other eyes than those of positive science. Philosophy, then, invades the domain of experience. She busies herself with many things which hitherto have not concerned her. Science, theory of knowledge, and metaphysics find themselves on the same ground. At first there may be a certain confusion. All three may think they have lost something. But all three will profit from the meeting.

Positive science, indeed, may pride itself on the uniform value attributed to its affirmations in the whole field of experience. But, if they are all placed on the same footing, they are all tainted with the same relativity. It is not so, if we begin by making the distinction which, in our view, is forced upon us. The understanding is at home in the domain of unorganized matter. On this matter human action is naturally exercised; and action, as we said above, cannot be set in motion in the unreal. Thus, of physics—so long as we are considering only its general form and not the particular cutting out of matter in which it is manifested—we may say that it touches the absolute. On the contrary, it is by accident—chance or convention, as you please—that science obtains a hold on the living analogous to the hold it has on matter. Here the use of conceptual frames is no longer natural. I do not wish to say that it is not legitimate, in the scientific meaning of the term. If science is to extend our action on things, and if we can act only with inert matter for instrument, science can and must continue to treat the living as it has treated the inert. But, in doing so, it must be understood that the further it penetrates the depths of *life*, the more symbolic, the more relative to the contingencies of action, the knowledge it supplies to us becomes. On this new ground philosophy ought then to follow science, in order to superpose on scientific truth a knowledge of another kind, which may be called metaphysical. Thus combined, all our knowledge, both scientific and metaphysical, is heightened. In the absolute we live and move and have our being. The knowledge we possess of it is incomplete, no doubt, but not external or relative. It is reality itself, in the profoundest meaning of the word, that we reach by the combined and progressive development of science and of philosophy.

Thus, in renouncing the factitious unity which the understanding imposes on nature from outside, we shall perhaps find its true, inward and living unity. For the effort we make to transcend the pure understanding introduces us into that more vast something out of which our understanding is cut, and from which it has detached itself. And, as matter is determined by intelligence, as there is between them an evident agreement, we cannot make the genesis of the one without making the genesis of the other. An identical process must have cut out matter and the intellect, at the same time, from a stuff that contained both. Into this reality we shall get back more and more completely, in proportion as we compel ourselves to transcend pure intelligence.

Let us then concentrate attention on that which we have that is at the same time the most removed from externality and the least penetrated with intellectuality. Let us seek, in the depths of our experience, the point where we feel ourselves most intimately within our own life. It is into pure duration that we then plunge back, a duration in which the past, always moving on, is swelling unceasingly with a present that is absolutely new. But, at the same time, we feel the spring of our will strained to its utmost limit. We must, by a strong recoil of our personality on itself, gather up our past which is slipping away, in order to thrust it, compact and undivided, into a present which it will create by entering. Rare indeed are the moments when we are self-possessed to this extent: it is then that our actions are truly free. And even at these moments we do not completely possess ourselves. Our feeling of duration, I should say the actual coinciding of ourself with itself, admits of degrees. But the more the feeling is deep and the coincidence complete, the more the life in which it replaces us absorbs intellectuality by transcending it. For the natural function of the intellect is to bind like to like, and it is only facts that can be repeated that are entirely adaptable to intellectual conceptions. Now, our intellect does undoubtedly grasp the real moments of real duration after they are past; we do so by reconstituting the new state of consciousness out of a series of views taken of it from the outside, each of which resembles as much as possible something already known; in this sense we may say that the state of consciousness contains intellectuality implicitly. Yet the state of consciousness overflows the intellect; it is indeed incommensurable with the intellect, being itself indivisible and new.

Now let us relax the strain, let us interrupt the effort to crowd as much as possible of the past into the present. If the relaxation were complete, there would no longer be either memory or will—which amounts to saying that, in fact, we never do fall into this absolute passivity, any more than we can make ourselves absolutely free. But, in the limit, we get a glimpse of an existence made of a present which recommences unceasingly—devoid of real duration, nothing but the instantaneous which dies and is born again endlessly. Is the existence of matter of this nature? Not altogether, for analysis resolves it into elementary vibrations, the shortest of which are of very slight duration, almost vanishing, but not nothing. It may be presumed, nevertheless, that physical existence inclines in this second direction, as psychical existence in the first.

Behind "spirituality" on the one hand, and "materiality" with intellectuality on the other, there are then two processes opposite in their direction, and we pass from the first to the second by way of inversion, or perhaps even by simple interruption, if it is true that inversion and interruption are two terms which in this case must be held to be synonymous, as we shall show at more length later on. This presumption is confirmed when we consider things from the point of view of extension, and no longer from that of duration alone.

The more we succeed in making ourselves conscious of our progress in pure duration, the more we feel the different parts of our being enter into each other, and our whole personality concentrate itself in a point, or rather a sharp edge, pressed against the future and cutting into it unceasingly. It is in this that life and action are free. But suppose we let ourselves go and, instead of acting, dream. At once the self is scattered; our past, which till then was gathered together into the indivisible impulsion it communicated to us, is broken up into a thousand recollections made external to one another. They give up interpenetrating in the degree that they become fixed. Our personality thus descends in the direction of space. It coasts around it continually in sensation. We will not dwell here on a point we have studied elsewhere. Let us merely recall that extension admits of degrees, that all sensation is extensive in a certain measure, and that the idea of unextended sensations, artificially localized in space, is a mere view of the mind, suggested by an unconscious metaphysic much more than by psychological observation.

No doubt we make only the first steps in the direction of the extended, even when we let ourselves go as much as we can. But suppose for a moment that matter consists in this very movement pushed further, and that physics is simply psychics inverted. We shall now understand why the mind feels at its ease, moves about naturally in space, when matter suggests the more distinct idea of it. This space is already possessed as an implicit idea in its own eventual *detension*, that is to say, of its own possible *extension*. The mind finds space in things, but could have got it without them if it had had imagination strong enough to push the inversion of its own natural movement to the end. On the other hand, we are able to explain how matter accentuates still more its materiality, when viewed by the mind. Matter, at first, aided mind to run down its own incline; it gave the impulsion. But, the impulsion once received, mind continues its course. The idea that it forms of *pure* space is only the *schema* of the limit at which this movement would end. Once in possession of the form of space, mind uses it like a net with meshes that can be made and unmade at will, which, thrown over matter, divides it as the needs of our action demand. Thus, the space of our geometry and the spatiality of things are mutually engendered by the reciprocal action and reaction of two terms which are essentially the same, but which move each in the direction inverse of the other. Neither is space so foreign to our nature as we imagine, nor is matter as completely extended in space as our senses and intellect represent it.

We have treated of the first point elsewhere. As to the second, we will limit ourselves to pointing out that perfect spatiality would consist in a perfect externality of parts in their relation to one another, that is to say, in a complete reciprocal independence. Now, there is no material point that does not act on every other material point. When we observe that a thing really *is* there where it *acts*, we shall be led to say (as Faraday[1] was) that all the atoms interpenetrate and that each of them fills the world. On such a hypothesis, the atom or, more generally, the material point, becomes simply a view of the mind, a view which we come to take when we continue far enough the work (wholly relative to our faculty of acting) by which we subdivide matter into bodies. Yet it is undeniable that matter lends itself to this subdivision, and that, in supposing it breakable into parts external to one another, we are constructing a science sufficiently representative of the real. It is undeniable that if there be no entirely isolated system, yet science finds means of cutting up the universe into systems relatively independent of each other, and commits no appreciable error in

[1] Faraday, *A Speculation concerning Electric Conduction* (*Philosophical Magazine*, 3d. series, vol. xxiv.).

doing so. What else can this mean but that matter *extends* itself in space without being absolutely *extended* therein, and that in regarding matter as decomposable into isolated systems, in attributing to it quite distinct elements which change in relation to each other without changing in themselves (which are "displaced," shall we say, without being "altered"), in short, in conferring on matter the properties of pure space, we are transporting ourselves to the terminal point of the movement of which matter simply indicates the direction?

What the *Transcendental Aesthetic* of Kant appears to have established once for all is that extension is not a material attribute of the same kind as others. We cannot reason indefinitely on the notions of heat, color, or weight: in order to know the modalities of weight or of heat, we must have recourse to experience. Not so of the notion of space. Supposing even that it is given empirically by sight and touch (and Kant has not questioned the fact) there is this about it that is remarkable that our mind, speculating on it with its own powers alone, cuts out in it, *a priori*, figures whose properties we determine *a priori*: experience, with which we have not kept in touch, yet follows us through the infinite complications of our reasonings and invariably justifies them. That is the fact. Kant has set it in clear light. But the explanation of the fact, we believe, must be sought in a different direction to that which Kant followed.

Intelligence, as Kant represents it to us, is bathed in an atmosphere of spatiality to which it is as inseparably united as the living body to the air it breathes. Our perceptions reach us only after having passed through this atmosphere. They have been impregnated in advance by our geometry, so that our faculty of thinking only finds again in matter the mathematical properties which our faculty of perceiving has already deposed there. We are assured, therefore, of seeing matter yield itself with docility to our reasonings; but this matter, in all that it has that is intelligible, is our own work; of the reality "in itself" we know nothing and never shall know anything, since we only get its refraction through the forms of our faculty of perceiving. So that if we claim to affirm something of it, at once there rises the contrary affirmation, equally demonstrable, equally plausible. The ideality of space is proved directly by the analysis of knowledge indirectly by the antinomies to which the opposite theory leads. Such is the governing idea of the Kantian criticism. It has inspired Kant with a peremptory refutation of "empiricist" theories of knowledge. It is, in our opinion, definitive in what it denies. But, in what it affirms, does it give us the solution of the problem?

With Kant, space is given as a ready-made form of our perceptive faculty—a veritable *deus ex machina*, of which we see neither how it arises, nor why it is what it is rather than anything else. "Things-in-themselves" are also given, of which he claims that we can know nothing: by what right, then, can he affirm their existence, even as "problematic"? If the unknowable reality projects into our perceptive faculty a "sensuous manifold" capable of fitting into it exactly, is it not, by that very fact, in part known? And when we examine this exact fitting, shall we not be led, in one point at least, to suppose a pre-established harmony between things and our mind—an idle hypothesis, which Kant was right in wishing to avoid? At bottom, it is for not having distinguished degrees in spatiality that he has had to take space ready-made as given—whence the question how the "sensuous manifold" is adapted to it. It is for the same reason that he has supposed matter wholly developed into parts absolutely external to one another;—whence antinomies, of which we may plainly see that the thesis and antithesis suppose the perfect coincidence of matter with geometrical space, but which vanish the moment we cease to extend to matter what is true only of pure space. Whence, finally, the conclusion that there are three alternatives, and three only, among which to choose a theory of knowledge: either the mind is determined by things, or things are determined by the mind, or between mind and things we must suppose a mysterious agreement.

But the truth is that there is a fourth, which does not seem to have occurred to Kant—in the first place because he did not think that the mind overflowed the intellect, and in the second place (and this is at bottom the same thing) because he did not attribute to duration an absolute existence, having put time, *a priori*, on the same plane as space. This alternative consists, first of all, in regarding the intellect as a special function of the mind, essentially turned toward inert matter; then in saying that neither does matter determine the form of the intellect, nor does the intellect impose its form on matter nor have matter and intellect been regulated in regard to one another by we know not what pre-established harmony, but that intellect and matter have progressively adapted themselves one to the other in order to attain at last a common form. *This adaptation has, moreover, been brought about quite naturally, because it is the same in-*

version of the same movement which creates at once the intellectuality of mind and the materiality of things.

From this point of view the knowledge of matter that our perception on one hand and science on the other give to us appears, no doubt, as approximative, but not as relative. Our perception, whose rôle it is to hold up a light to our actions, works a dividing up of matter that is always too sharply defined, always subordinated to practical needs, consequently always requiring revision. Our science, which aspires to the mathematical form, overaccentuates the spatiality of matter; its formulae are, in general, too precise, and ever need remaking. For a scientific theory to be final, the mind would have to embrace the totality of things in block and place each thing in its exact relation to every other thing; but in reality we are obliged to consider problems one by one, in terms which are, for that very reason, provisional, so that the solution of each problem will have to be corrected indefinitely by the solution that will be given to the problems that will follow: thus, science as a whole is relative to the particular order in which the problems happen to have been put. It is in this meaning, and to this degree, that science must be regarded as conventional. But it is a conventionality of fact so to speak, and not of right. In principle, positive science bears on reality itself, provided it does not overstep the limits of its own domain, which is inert matter.

Scientific knowledge, thus regarded, rises to a higher plane. In return, the theory of knowledge becomes an infinitely difficult enterprise, and which passes the powers of the intellect alone. It is not enough to determine, by careful analysis, the categories of thought; we must engender them. As regards space, we must, by an effort of mind *sui generis,* follow the progression or rather the regression of the extra-spatial degrading itself into spatiality. When we make ourselves self-conscious in the highest possible degree and then let ourselves fall back little by little, we get the feeling of extension: we have an extension of the self into recollections that are fixed and external to one another, in place of the tension it possessed as an indivisible active will. But this is only a beginning. Our consciousness, sketching the movement, shows us its direction and reveals to us the possibility of continuing it to the end; but consciousness itself does not go so far. Now, on the other hand, if we consider matter, which seems to us at first coincident with space, we find that the more our attention is fixed on it, the more the parts which we said were laid side by side enter into each other, each of them undergoing the action of the whole, which is consequently somehow present in it. Thus, although matter stretches itself out in the direction of space, it does not completely attain it; whence we may conclude that it only carries very much further the movement that consciousness is able to sketch within us in its nascent state. We hold, therefore, the two ends of the chain, though we do not succeed in seizing the intermediate links. Will they always escape us? We must remember that philosophy, as we define it, has not yet become completely conscious of itself. Physics understands its rôle when it pushes matter in the direction of spatiality; but has metaphysics understood its rôle when it has simply trodden in the steps of physics, in the chimerical hope of going further in the same direction? Should not its own task be, on the contrary, to remount the incline that physics descends, to bring back matter to its origins, and to build up progressively a cosmology which would be, so to speak, a reversed psychology? All that which seems *positive* to the physicist and to the geometrician would become, from this new point of view, an interruption or inversion of the true positivity, which would have to be defined in psychological terms.

When we consider the admirable order of mathematics, the perfect agreement of the objects it deals with, the immanent logic in numbers and figures, our certainty of always getting the same conclusion, however diverse and complex our reasonings on the same subject, we hesitate to see in properties apparently so positive a system of negations, the absence rather than the presence of a true reality. But we must not forget that our intellect, which finds this order and wonders at it, is directed in the same line of movement that leads to the materiality and spatiality of its object. The more complexity the intellect puts into its object by analyzing it, the more complex is the order it finds there. And this order and this complexity necessarily appear to the intellect as a positive reality, since reality and intellectuality are turned in the same direction.

When a poet reads me his verses, I can interest myself enough in him to enter into his thought, put myself into his feelings, live over again the simple state he has broken into phrases and words. I sympathize then with his inspiration, I follow it with a continuous movement which is, like the inspiration itself, an undivided act. Now, I need only relax my attention, let go the tension that there is in me, for the sounds, hitherto swallowed up in the sense, to appear to me distinctly, one by one, in their materiality. For this I have not to do anything; it is enough to withdraw something. In proportion as I let myself go, the successive sounds will become the more individualized; as the phrases were broken into words, so the words will scan in syllables which I shall perceive one after another. Let me go farther still in the direction of dream: the letters themselves will become loose and will be seen to dance along, hand in hand, on some fantastic sheet of paper. I shall then admire the precision of the interweavings, the marvelous order of the procession, the exact insertion of the letters into the syllables, of the syllables into the words and of the words into the sentences. The farther I pursue this quite negative direction of relaxation, the more extension and complexity I shall create; and the more the complexity in its turn increases, the more admirable will seem to be the order which continues to reign, undisturbed, among the elements. Yet this complexity and extension represent nothing positive; they express a deficiency of will. And, on the other hand, the order must grow with the complexity, since it is only an aspect of it. The more we perceive, symbolically, parts in an indivisible whole, the more the number of the relations that the parts have between themselves necessarily increases, since the same undividedness of the real whole continues to hover over the growing multiplicity of the symbolic elements into which the scattering of the attention has decomposed it. A comparison of this kind will enable us to understand, in some measure, how the same suppression of positive reality, the same inversion of a certain original movement, can create at once extension in space and the admirable order which mathematics finds there. There is, of course, this difference between the two cases, that words and letters have been invented by a positive effort of humanity, while space arises automatically, as the remainder of a subtraction arises once the two numbers are posited.[1] But, in the one case as in the other, the infinite complexity of the parts and their perfect co-ordination among themselves are created at one and the same time by an inversion which is, at bottom, an interruption, that is to say, a diminution of positive reality.

All the operations of our intellect tend to geometry, as to the goal where they find their perfect fulfillment. But, as geometry is necessarily prior to them (since these operations have not as their end to construct space and cannot do otherwise than take it as given) it is evident that it is a latent geometry, immanent in our idea of space, which is the main spring of our intellect and the

[1] Our comparison does no more than develop the content of the term λόγος, as Plotinus understands it. For while the λόγος of this philosopher is a generating and informing power, an aspect or a fragment of the ψυχή, on the other hand Plotinus sometimes speaks of it as of a *discourse*. More generally, the relation that we establish in the present chapter between "extension" and "detension" resembles in some aspects that which Plotinus supposes (some developments of which must have inspired M. Ravaisson) when he makes extension not indeed an inversion of original Being, but an enfeeblement of its essence, one of the last stages of the procession (see in particular, *Enn.* IV. iii. 9-11, and III. vi. 17-18). Yet ancient philosophy did not see what consequences would result from this for mathematics, for Plotinus, like Plato, erected mathematical essences into absolute realities. Above all, it suffered itself to be deceived by the purely superficial analogy of duration with extension. It treated the one as it treated the other, regarding change as a degradation of immutability, the sensible as a fall from the intelligible. Whence, as we shall show in the next chapter, a philosophy which fails to recognize the real function and scope of the intellect.

cause of its working. We shall be convinced of this if we consider the two essential functions of intellect, the faculty of deduction and that of induction.

Let us begin with deduction. The same movement by which I trace a figure in space engenders its properties: they are visible and tangible in the movement itself; I feel, I see in space the relation of the definition to its consequences, of the premises to the conclusion. All the other concepts of which experience suggests the idea to me are only in part constructible *a priori*; the definition of them is therefore imperfect, and the deductions into which these concepts enter, however closely the conclusion is linked to the premises, participate in this imperfection. But when I trace roughly in the sand the base of a triangle, as I begin to form the two angles at the base, I know positively, and understand absolutely, that if these two angles are equal the sides will be equal also, the figure being then able to be turned over on itself without there being any change whatever. I know it before I have learnt geometry. Thus, prior to the science of geometry, there is a natural geometry whose clearness and evidence surpass the clearness and evidence of other deductions. Now, these other deductions bear on qualities, and not on magnitudes purely. They are, then, likely to have been formed on the model of the first, and to borrow their force from the fact that, behind quality, we see magnitude vaguely showing through. We may notice, as a fact, that questions of situation and of magnitude are the first that present themselves to our activity, those which intelligence externalized in action resolves even before reflective intelligence has appeared. The savage understands better than the civilized man how to judge distances, to determine a direction, to retrace by memory the often complicated plan of the road he has traveled, and so to return in a straight line to his starting-point.[1] If the animal does not deduce explicitly, if he does not form explicit concepts, neither does he form the idea of a homogeneous space. You cannot present this space to yourself without introducing, in the same act, a virtual geometry which will, of itself, degrade itself into logic. All the repugnance that philosophers manifest toward this manner of regarding things comes from this, that the logical work of the intellect represents to their eyes a positive spiritual effort. But, if we understand by spirituality a progress to ever new creations, to conclusions incommensurable with the premises and indeterminable by relation to them, we must say of an idea that moves among relations of necessary determination, through premises which contain their conclusion in advance, that it follows the inverse direction, that of materiality. What appears, from the point of view of the intellect, as an effort, is in itself a letting go. And while, from the point of view of the intellect, there is a *petitio principii* in making geometry arise automatically from space, and logic from geometry—on the contrary, if space is the ultimate goal of the mind's movement of *detension*, space cannot be given without positing also logic and geometry, which are along the course of the movement of which pure spatial intuition is the goal.

It has not been enough noticed how feeble is the reach of deduction in the psychological and moral sciences. From a proposition verified by facts, verifiable consequences can here be drawn only up to a certain point,

[1] Bastian, *The Brain as an Organ of the Mind*, pp. 214-16.

only in a certain measure. Very soon appeal has to be made to common sense, that is to say, to the continuous experience of the real, in order to inflect the consequences deduced and bend them along the sinuosities of life. Deduction succeeds in things moral only metaphorically, so to speak, and just in the measure in which the moral is transposable into the physical, I should say translatable into spatial symbols. The metaphor never goes very far, any more than a curve can long be confused with its tangent. Must we not be struck by this feebleness of deduction as something very strange and even paradoxical? Here is a pure operation of the mind, accomplished solely by the power of the mind. It seems that, if anywhere it should feel at home and evolve at ease, it would be among the things of the mind, in the domain of the mind. Not at all; it is there that it is immediately at the end of its tether. On the contrary, in geometry, in astronomy, in physics, where we have to do with things external to us, deduction is all-powerful! Observation and experience are undoubtedly necessary in these sciences to arrive at the principle, that is, to discover the aspect under which things must be regarded; but, strictly speaking, we might, by good luck, have hit upon it at once; and, as soon as we possess this principle, we may draw from it, at any length, consequences which experience will always verify. Must we not conclude, therefore, that deduction is an operation governed by the properties of matter, molded on the mobile articulations of matter, implicitly given, in fact, with the space that underlies matter? As long as it turns upon space or spatialized time, it has only to let itself go. It is *duration* that puts spokes in its wheels.

Deduction, then, does not work unless there be spatial intuition behind it. But we may say the same of induction. It is not necessary indeed to think geometrically, nor even to think at all, in order to expect from the same conditions a repetition of the same fact. The consciousness of the animal already does this work, and indeed, independently of all consciousness, the living body itself is so constructed that it can extract from the successive situations in which it finds itself the similarities which interest it, and so respond to the stimuli by appropriate reactions. But it is a far cry from a mechanical expectation and reaction of the body, to induction properly so called, which is an intellectual operation. Induction rests on the belief that there are causes and effects, and that the same effects follow the same causes. Now, if we examine this double belief, this is what we find. It implies, in the first place, that reality is decomposable into groups, which can be practically regarded as isolated and independent. If I boil water in a kettle on a stove, the operation and the objects that support it are, in reality, bound up with a multitude of other objects and a multitude of other operations; in the end, I should find that our entire solar system is concerned in what is being done at this particular point of space. But, in a certain measure, and for the special end I am pursuing, I may admit that things happen as if the group *water-kettle-stove* were an independent microcosm. That is my first affirmation. Now, when I say that this microcosm will always behave in the same way, that the heat will necessarily, at the end of a certain time, cause the boiling of the water, I admit that it is sufficient that a certain number of elements of the system be given in order that the system should be complete; it completes itself automatically, I am not free to complete it in thought as I please. The stove, the kettle and the water being given, with a certain interval of duration, it seems to me that the boiling, which experience showed me yesterday to be the only thing wanting to complete the system, will complete it tomorrow, no matter when tomorrow may be. What is there at the base of this belief? Notice that the belief is more or less assured, according as the case may be, but that it is forced upon the mind as an absolute necessity when the microcosm considered contains only magnitudes. If two numbers be given, I am not free to choose their difference. If two sides of a triangle and the contained angle are given, the third side arises of itself and the triangle completes itself automatically. I can, it matters not where and it matters not when, trace the same two sides containing the same angle: it is evident that the new triangles so formed can be superposed on the first, and that consequently the same third side will come to complete the system. Now, if my certitude is perfect in the case in which I reason on pure space determinations, must I not suppose that, in the other cases, the certitude is greater the nearer it approaches this extreme case? Indeed, may it not be the limiting case which is seen through all the others and which colors them, accordingly as they are more or less transparent, with a more or less pronounced tinge of geometrical necessity? [1] In fact, when I say that the water on the fire will boil today as it did yesterday, and that this is an absolute necessity, I feel vaguely that my imagination is placing

[1] We have dwelt on this point in a former work. See the *Essai sur les données immédiates de la conscience*. Paris, 1889, pp. 155-160.

the stove of yesterday on that of today, kettle on kettle, water on water, duration on duration, and it seems then that the rest must coincide also, for the same reason that, when two triangles are superposed and two of their sides coincide, their third sides coincide also. But my imagination acts thus only because it shuts its eyes to two essential points. For the system of today actually to be superimposed on that of yesterday, the latter must have waited for the former, time must have halted, and everything become simultaneous: that happens in geometry, but in geometry alone. Induction therefore implies first that, in the world of the physicist as in that of the geometrician, time does not count. But it implies also that qualities can be superposed on each other like magnitudes. If, in imagination, I place the stove and fire of today on that of yesterday, I find indeed that the form has remained the same; it suffices, for that, that the surfaces and edges coincide; but what is the coincidence of two qualities, and how can they be superposed one on another in order to ensure that they are identical? Yet I extend to the second order of reality all that applies to the first. The physicist legitimates this operation later on by reducing, as far as possible, differences of quality to differences of magnitude; but, prior to all science, I incline to liken qualities to quantities, as if I perceived behind the qualities, as through a transparency, a geometrical mechanism.[1] The more complete this transparency, the more it seems to me that in the same conditions there must be a repetition of the same fact. Our inductions are certain, to our eyes, in the exact degree in which we make the qualitative differences melt into the

[1] *Op. cit.* chaps. i. and ii. *passim.*

homogeneity of the space which subtends them, so that geometry is the ideal limit of our inductions as well as of our deductions. The movement at the end of which is spatiality lays down along its course the faculty of induction as well as that of deduction, in fact, intellectuality entire.

It creates them in the mind. But it creates also, in things, the "order" which our induction, aided by deduction, finds there. This order, on which our action leans and in which our intellect recognizes itself, seems to us marvelous. Not only do the same general causes always produce the same general effects, but beneath the visible causes and effects our science discovers an infinity of infinitesimal changes which work more and more exactly into one another, the further we push the analysis: so much so that, at the end of this analysis, matter becomes, it seems to us, geometry itself. Certainly, the intellect is right in admiring here the growing order in the growing complexity; both the one and the other must have a positive reality for it, since it looks upon itself as positive. But things change their aspect when we consider the whole of reality as an undivided advance forward to successive creations. It seems to us, then, that the complexity of the material elements and the mathematical order that binds them together must arise automatically when within the whole a partial interruption or inversion is produced. Moreover, as the intellect itself is cut out of mind by a process of the same kind, it is attuned to this order and complexity, and admires them because it recognizes itself in them. But what is admirable *in itself*, what really deserves to provoke wonder, is the ever-renewed creation which reality, whole and undivided, accomplishes in advancing; for no complication of the mathematical order with itself, however elaborate we may suppose it, can introduce an atom of novelty into the world, whereas this power of creation once given (and it exists, for we are conscious of it in ourselves, at least when we act freely) has only to be diverted from itself to relax its tension, only to relax its tension to extend, only to extend for the mathematical order of the elements so distinguished and the inflexible determinism connecting them to manifest the interruption of the creative act: in fact, inflexible determinism and mathematical order are one with this very interruption.

It is this merely negative tendency that the particular laws of the physical world express. None of them, taken separately, has objective reality; each is the work of an investigator who has regarded things from a certain bias, isolated certain variables, applied certain conventional units of measurement. And yet there is an order approximately mathematical immanent in matter, an objective order, which our science approaches in proportion to its progress. For if matter is a relaxation of the inextensive into the extensive and, thereby, of liberty into necessity, it does not indeed wholly coincide with pure homogeneous space, yet is constituted by the movement which leads to space, and is therefore on the way to geometry. It is true that laws of mathematical form will never apply to it completely. For that, it would have to be pure space and step out of duration.

We cannot insist too strongly that there is something artificial in the mathematical form of a physical law, and consequently in our scientific knowledge of things.[1] Our standards of measurement are conventional, and, so to say, foreign to the intentions of nature: can we suppose that nature has related all the modalities of heat to the expansion of the same mass of mercury, or to the change of pressure of the same mass of air kept at a constant volume? But we may go further. In a general way, *measuring* is a wholly human operation, which implies that we really or ideally superpose two objects one on another a certain number of times. Nature did not dream of this superposition. It does not measure, nor does it count. Yet physics counts, measures, relates "quantitative" variations to one another to obtain laws, and it succeeds. Its success would be inexplicable, if the movement which constitutes materiality were not the same movement which, prolonged by us to its end, that is to say, to homogeneous space, results in making us count, measure, follow in their respective variations terms that are functions one of another. To effect this prolongation of the movement, our intellect has only to let itself go, for it runs naturally to space and mathematics, intellectuality and materiality being of the same nature and having been produced in the same way.

If the mathematical order were a positive thing, if there were, immanent in matter, laws comparable to those of our codes, the success of our science would have in it something of the miraculous. What chances should we have indeed of finding the standard of nature and of isolating exactly, in order to determine their reciprocal

[1] Cf. especially the profound studies of M. Ed. Le Roy in the *Revue de métaph. et de morale.*

relations, the very variables which nature has chosen? But the success of a science of mathematical form would be no less incomprehensible, if matter did not already possess everything necessary to adapt itself to our formulae. One hypothesis only, therefore, remains plausible, namely, that the mathematical order is nothing positive, that it is the form toward which a certain *interruption* tends of itself, and that materiality consists precisely in an interruption of this kind. We shall understand then why our science is contingent, relative to the variables it has chosen, relative to the order in which it has successively put the problems, and why nevertheless it succeeds. It might have been, as a whole, altogether different, and yet have succeeded. This is so, just because there is no definite system of mathematical laws, at the base of nature, and because mathematics in general represents simply the side to which matter inclines. Put one of those little cork dolls with leaden feet in any posture, lay it on its back, turn it up on its head, throw it into the air: it will always stand itself up again, automatically. So likewise with matter: we can take it by any end and handle it in any way, it will always fall back into some one of our mathematical formulae, because it is weighted with geometry.

But the philosopher will perhaps refuse to found a theory of knowledge on such considerations. They will be repugnant to him, because the mathematical order, being order, will appear to him to contain something positive. It is in vain that we assert that this order produces itself automatically by the interruption of the inverse order, that it is this very interruption. The idea persists, none the less, that *there might be no order at all*, and that the mathematical order of things, being a conquest over disorder, possesses a positive reality. In examining this point, we shall see what a prominent part the idea of *disorder* plays in problems relative to the theory of knowledge. It does not appear explicitly, and that is why it escapes our attention. It is, however, with the criticism of this idea that a theory of knowledge ought to begin, for if the great problem is to know why and how reality submits itself to an order, it is because the absence of every kind of order appears possible or conceivable. It is this absence of order that realists and idealists alike believe they are thinking of—the realist when he speaks of the regularity that "objective" laws actually impose on a virtual disorder of nature, the idealist when he supposes a "sensuous manifold" which is coordinated (and consequently itself without order) under the organizing influence of our understanding. The idea of disorder, in the sense of *absence of order,* is then what must be analyzed first. Philosophy borrows it from daily life. And it is unquestionable that, when ordinarily we speak of disorder, we are thinking of something. But of what?

It will be seen in the next chapter how hard it is to determine the content of a negative idea, and what illusions one is liable to, what hopeless difficulties philosophy falls into, for not having undertaken this task. Difficulties and illusions are generally due to this, that we accept as final a manner of expression essentially provisional. They are due to our bringing into the domain of speculation a procedure made for practice. If I choose a volume in my library at random, I may put it back on the shelf after glancing at it and say, "This is not verse." Is this what I have really seen in turning over the leaves of the book? Obviously not. I have not seen, I never shall see, an absence of verse. I have seen prose. But as it is poetry I want, I express what I find as a function of what I am looking for, and instead of saying, "This is prose," I say, "This is not verse." In the same way, if the fancy takes me to read prose, and I happen on a volume of verse, I shall say, "This is not prose," thus expressing the data of my perception, which shows me verse, in the language of my expectation and attention, which are fixed on the idea of prose and will hear of nothing else. Now, if M. Jourdain heard me, he would infer, no doubt, from my two exclamations that prose and poetry are two forms of language reserved for books, and that these learned forms have come and overlaid a language which was neither prose nor verse. Speaking of this thing which is neither verse nor prose, he would suppose, moreover, that he was thinking of it: it would be only a pseudo-idea, however. Let us go further still: the pseudo-idea would create a pseudo-problem, if M. Jourdain were to ask his professor of philosophy how the prose form and the poetry form have been superadded to that which possessed neither the one nor the other, and if he wished the professor to construct a theory of the imposition of these two forms upon this formless matter. His question would be absurd, and the absurdity would lie in this, that he was hypostasizing as the substratum of prose and poetry the simultaneous negation of both, forgetting that the negation of the one consists in the affirmation of the other.

Now, suppose that there are two species of order, and

that these two orders are two contraries within one and the same genus. Suppose also that the idea of disorder arises in our mind whenever, seeking one of the two kinds of order, we find the other. The idea of disorder would then have a clear meaning in the current practice of life: it would objectify, for the convenience of language, the disappointment of a mind that finds before it an order different from what it wants, an order with which it is not concerned at the moment, and which, in this sense, does not exist for it. But the idea would not admit a theoretical use. So if we claim, notwithstanding, to introduce it into philosophy, we shall inevitably lose sight of its true meaning. It denotes the absence of a certain order, but *to the profit of another* (with which we are not concerned); only, as it applies to each of the two in turn, and as it even goes and comes continually between the two, we take it on the way, or rather on the wing, like a shuttlecock between two battledores, and treat it as if it represented, not the absence of the one or other order as the case may be, but the absence of both together—a thing that is neither perceived nor conceived, a simple verbal entity. So there arises the problem how order is imposed on disorder, form on matter. In analyzing the idea of disorder thus subtilized, we shall see that it represents nothing at all, and at the same time the problems that have been raised around it will vanish.

It is true that we must begin by distinguishing, and even by opposing one to the other, two kinds of order which we generally confuse. As this confusion has created the principal difficulties of the problem of knowledge, it will not be useless to dwell once more on the marks by which the two orders are distinguished.

In a general way, reality is *ordered* exactly to the degree in which it satisfies our thought. Order is therefore a certain agreement between subject and object. It is the mind finding itself again in things. But the mind, we said, can go in two opposite ways. Sometimes it follows its natural direction: there is then progress in the form of tension, continuous creation, free activity. Sometimes it inverts it, and this inversion, pushed to the end, leads to extension, to the necessary reciprocal determination of elements externalized each by relation to the others, in short, to geometrical mechanism. Now, whether experience seems to us to adopt the first direction or whether it is drawn in the direction of the second, in both cases we say there is order, for in the two processes the mind finds itself again. The confusion between them is therefore natural. To escape it, different names would have to be given to the two kinds of order, and that is not easy, because of the variety and variability of the forms they take. The order of the second kind may be defined as geometry, which is its extreme limit; more generally, it is that kind of order that is concerned whenever a relation of necessary determination is found between causes and effects. It evokes ideas of inertia, of passivity, of automatism. As to the first kind of order, it oscillates no doubt around finality; and yet we cannot define it as finality, for it is sometimes above, sometimes below. In its highest forms, it is more than finality, for of a free action or a work of art we may say that they show a perfect order, and yet they can only be expressed in terms of ideas approximately, and after the event. Life in its entirety, regarded as a creative evolution, is something analogous; it transcends finality, if we understand by finality the realization of an idea conceived or conceivable in advance. The category of finality is therefore too narrow for life in its entirety. It is, on the other hand, often too wide for a particular manifestation of life taken separately. Be that as it may, it is with the *vital* that we have here to do, and the whole present study strives to prove that the vital is in the direction of the voluntary. We may say then that this first kind of order is that of the *vital* or of the *willed*, in opposition to the second, which is that of the *inert* and the *automatic*. Common sense instinctively distinguishes between the two kinds of order, at least in the extreme cases; instinctively, also, it brings them together. We say of astronomical phenomena that they manifest an admirable order, meaning by this that they can be foreseen mathematically. And we find an order no less admirable in a symphony of Beethoven, which is genius, originality, and therefore unforeseeability itself.

But it is exceptional for order of the first kind to take so distinct a form. Ordinarily, it presents features that we have every interest in confusing with those of the opposite order. It is quite certain, for instance, that if we could view the evolution of life in its entirety, the spontaneity of its movement and the unforeseeability of its procedures would thrust themselves on our attention. But what we meet in our daily experience is a certain determinate living being, certain special manifestations of life, which repeat, *almost*, forms and facts already known; indeed, the similarity of structure that we find everywhere between what generates and what is generated—a similarity that enables us to include any number of living individuals in the same group—is to our eyes the very type of the *generic*: the inorganic genera seem to us to take living genera as models. Thus the vital order, such as it is offered to us piecemeal in experience, presents the same character and performs the same function as the physical order: both cause experience to *repeat itself*, both enable our mind to *generalize*. In reality, this character has entirely different origins in the two cases, and even opposite meanings. In the second case, the type of this character, its ideal limit, as also its foundation, is the geometrical necessity in virtue of which the same components give the same resultant. In the first case, this character involves, on the contrary, the intervention of something which manages to obtain the same total effect although the infinitely complex elementary causes may be quite different. We insisted on this last point in our first chapter, when we showed how identical structures are to be met with on independent lines of evolution. But, without looking so far, we may presume that the reproduction only of the type of the ancestor by his descendants is an entirely different thing from the repetition of the same composition of forces which yields an identical resultant. When we think of the infinity of infinitesimal elements and of infinitesimal causes that concur in the genesis of a living being, when we reflect that the absence or the deviation of one of them would spoil everything, the first impulse of the mind is to consider this army of little workers as watched over by a skilled foreman, the "vital principle," which is ever repairing faults, correcting effects of neglect or absentmindedness, putting things back in place: this is how we try to express the difference between the physical and the vital order, the former making the same combination of causes give the same combined effect, the latter securing the constancy of the effect even when there is some wavering in the causes. But that is only a comparison; on reflection, we find that there can be no foreman, for the very simple reason that there are no workers. The causes and elements that physico-chemical analysis discovers are real causes and elements, no doubt, as far as the facts of organic destruction are concerned; they are then limited in number. But vital phenomena, properly so called, or facts of organic creation open up to us, when we analyze them, the perspective of an analysis passing away to infinity: whence it may be inferred that the manifold causes and elements are here only views of the mind, attempting an ever closer and closer imitation of the operation of nature, while the operation imitated is an indivisible act. The likeness between individuals of the same species has thus an entirely different meaning, an entirely different origin, to that of the likeness between complex effects obtained by the same composition of the same causes. But in the one case as in the other, there is *likeness*, and consequently possible generalization. And as that is all that interests us in practice, since our daily life is and must be an expectation of the same things and the same situations, it is natural that this common character, essential from the point of view of our action, should bring the two orders together, in spite of a merely internal diversity between them which interests speculation only. Hence the idea of a *general order of nature*, everywhere the same, hovering over life and over matter alike. Hence our habit of designating by the same word and representing in the same way the existence of *laws* in the domain of inert matter and that of *genera* in the domain of life.

Now, it will be found that this confusion is the origin of most of the difficulties raised by the problem of knowledge, among the ancients as well as among the moderns. The generality of laws and that of genera having been designated by the same word and subsumed under the same idea, the geometrical order and the vital order are accordingly confused together. According to the point of view, the generality of laws is explained by that of genera, or that of genera by that of laws. The first view is characteristic of ancient thought; the second belongs to modern philosophy. But in both ancient and modern philosophy the idea of "generality" is an equivocal idea, uniting in its denotation and in its connotation incompatible objects and elements. In both there are grouped under the same concept two kinds of order which are alike only in the facility they give to our action on things. We bring together the two terms in virtue of a quite external likeness, which justifies no doubt their designation by the same word for practice, but which does not authorize us at all, in the speculative domain, to confuse them in the same definition.

The ancients, indeed, did not ask why nature submits to laws, but why it is ordered according to genera. The idea of genus corresponds more especially to an objective reality in the domain of life, where it expresses an unquestionable fact, heredity. Indeed, there can only be genera where there are individual objects; now, while the organized being is cut out from the general mass of matter by his very organization, that is to say naturally, it is our perception which cuts inert matter into distinct bodies. It is guided in this by the interests of action, by the nascent reactions that our body indicates—that is, as we have shown elsewhere,[1] by the potential genera that are trying to gain existence. In this, then, genera and individuals determine one another by a semi-artificial operation entirely relative to our future action on things. Nevertheless the ancients did not hesitate to put all genera in the same rank, to attribute the same absolute existence to all of them. Reality thus being a system of genera, it is to the generality of the genera (that is, in effect, to the generality expressive of the vital order) that the generality of laws itself had to be brought. It is interesting, in this respect, to compare the Aristotelian theory of the fall of bodies with the explanation furnished by Galileo. Aristotle is concerned solely with the concepts "high" and "low," "own proper place" as distinguished from "place occupied," "natural movement" and "forced movement";[2] the physical law in virtue of which the stone falls expresses for him that the stone regains the "natural place" of all stones, to wit, the earth. The stone, in his view, is not quite stone so long as it is not in its normal place; in falling back into this place it aims at completing itself, like a living being that grows, thus realizing fully the essence of the genus stone.[3] If this conception of the physical law were exact, the law would no longer be a mere relation established by the mind; the subdivision of matter into bodies would no

[1] *Matière et mémoire*, chapters iii. and iv.

[2] See in particular, *Phys.*, iv. 215 a 2; v. 230 b 12; viii. 225 a 2; and *De Caelo*, iv. 1-5; ii. 296 b 27; iv. 308 a 34.

[3] *De Caelo*, iv. 310 a 34 *τὸ δ'εἰς τὸν αὐτοῦ τόπον φέρεθαι ἕκαστον τὸ εἰς τὸ αὐτοῦ εἶδός ἐστι φέρεσθαι.*

longer be relative to our faculty of perceiving; all bodies would have the same individuality as living bodies, and the laws of the physical universe would express relations of real kinship between real genera. We know what kind of physics grew out of this, and how, for having believed in a science unique and final, embracing the totality of the real and at one with the absolute, the ancients were confined, in fact, to a more or less clumsy interpretation of the physical in terms of the vital.

But there is the same confusion in the moderns, with this difference, however, that the relation between the two terms is inverted: laws are no longer reduced to genera, but genera to laws; and science, still supposed to be uniquely one, becomes altogether relative, instead of being, as the ancients wished, altogether at one with the absolute. A noteworthy fact is the eclipse of the problem of genera in modern philosophy. Our theory of knowledge turns almost entirely on the question of laws: genera are left to make shift with laws as best they can. The reason is, that modern philosophy has its point of departure in the great astronomical and physical discoveries of modern times. The laws of Kepler and of Galileo have remained for it the ideal and unique type of all knowledge. Now, a law is a relation between things or between facts. More precisely, a law of mathematical form expresses the fact that a certain magnitude is a function of one or several other variables appropriately chosen. Now, the choice of the variable magnitudes, the distribution of nature into objects and into facts, has already something of the contingent and the conventional. But, admitting that the choice is hinted at, if not prescribed, by experience, the law remains none the less a relation, and a relation is essentially a comparison; it has objective reality only for an intelligence that represents to itself several terms at the same time. This intelligence may be neither mine nor yours: a science which bears on laws may therefore be an objective science, which experience contains in advance and which we simply make it disgorge; but it is none the less true that a comparison of some kind must be effected here, impersonally if not by anyone in particular, and that an experience made of laws, that is, of terms *related* to other terms, is an experience made of comparisons, which, before we receive it, has already had to pass through an atmosphere of intellectuality. The idea of a science and of an experience entirely relative to the human understanding was therefore implicitly contained in the conception of a science one and integral, composed of laws: Kant only brought it to light. But this conception is the result of an arbitrary confusion between the generality of laws and that of genera. Though an intelligence be necessary to condition terms by relation to each other, we may conceive that in certain cases the terms themselves may exist independently. And if, beside relations of term to term, experience also presents to us independent terms, the living genera being something quite different from systems of laws, one half, at least, of our knowledge bears on the "thing-in-itself," the very reality. This knowledge may be very difficult, just because it no longer builds up its own object and is obliged, on the contrary, to submit to it; but, however little it cuts into its object, it is into the absolute itself that it bites. We may go further: the other half of knowledge is no longer so radically, so definitely relative as certain philosophers say, if we can establish that it bears on a reality of inverse order, a reality which we always express in mathematical laws, that is to say in relations that imply comparisons, but which lends itself to this work only because it is weighted with spatiality and consequently with geometry. Be that as it may, it is the confusion of two kinds of order that lies behind the relativism of the moderns, as it lay behind the dogmatism of the ancients.

We have said enough to mark the origin of this confusion. It is due to the fact that the "vital" order, which is essentially creation, is manifested to us less in its essence than in some of its accidents, those which *imitate* the physical and geometrical order; like it, they present to us repetitions that make generalization possible, and in that we have all that interests us. There is no doubt that life as a whole is an evolution, that is, an unceasing transformation. But life can progress only by means of the living, which are its depositaries. Innumerable living beings, almost alike, have to repeat each other in space and in time for the novelty they are working out to grow and mature. It is like a book that advances toward a new edition by going through thousands of reprints with thousands of copies. There is, however, this difference between the two cases, that the successive impressions are identical, as well as the simultaneous copies of the same impression, whereas representatives of one and the same species are never entirely the same, either in different points of space or at different moments of time. Heredity does not only transmit characters; it transmits also the impetus in virtue of which the characters are modified, and this impetus is vitality itself. That is why we say that the repetition which serves as the base of our generalizations is essential in the physical order, accidental in the vital order. The physical order is "automatic"; the vital order is, I will not say voluntary, but analogous to the order "willed."

Now, as soon as we have clearly distinguished between the order that is "willed" and the order that is "automatic," the ambiguity that underlies the idea of *disorder* is dissipated, and, with it, one of the principal difficulties of the problem of knowledge.

The main problem of the theory of knowledge is to know how science is possible, that is to say, in effect, why there is order and not disorder in things. That order exists is a *fact*. But, on the other hand, disorder, *which appears to us to be less than order*, is, it seems, of *right*. The existence of order is then a mystery to be cleared up, at any rate a problem to be solved. More simply, when we undertake to found order, we regard it as contingent, if not in things, at least as viewed by the mind: of a thing that we do not judge to be contingent we do not require an explanation. If order did not appear to us as a conquest over something, or as an addition to something (which something is thought to be the "absence of order"), ancient realism would not have spoken of a "matter" to which the Idea superadded itself, nor would modern idealism have supposed a "sensuous manifold" that the understanding organizes into nature. Now, it is unquestionable that all order is contingent, and conceived as such. But contingent in relation to what?

The reply, to our thinking, is not doubtful. An order is contingent, and seems so, in relation to the inverse order, as verse is contingent in relation to prose and prose in relation to verse. But, just as all speech which is not prose is verse and necessarily conceived as verse, just as all speech which is not verse is prose and necessarily conceived as prose, so any state of things that is not one of the two orders is the other and is necessarily conceived as the other. But it may happen that we do not realize what we are actually thinking of, and perceive the idea really present to our mind only through a mist of affective states. Anyone can be convinced of this by considering the use we make of the idea of disorder in daily life. When I enter a room and pronounce it to be "in disorder," what do I mean? The position of each object is explained by the automatic movements of the person who has slept in the room, or by the efficient causes, whatever they may be, that have caused each article of furniture, clothing, etc., to be where it is: the order, in the second sense of the word, is perfect. But it is order of the first kind that I am expecting, the order that a methodical person consciously puts into his life, the willed order and not the automatic: so I call the absence of this order "disorder." At bottom, all there is that is real, perceived and even conceived, in this absence of one of the two kinds of order, is the presence of the other. But the second is indifferent to me, *I am interested only in the first*, and I express the presence of the second as a function of the first, instead of expressing it, so to speak, as a function of itself, by saying it is *disorder*. Inversely, when we affirm that we are imagining a chaos, that is to say a state of things in which the physical world no longer obeys laws, what are we thinking of? We imagine facts that appear and disappear *capriciously*. First we think of the physical universe as we know it, with effects and causes well proportioned to each other; then, by a series of arbitrary decrees, we augment, diminish, suppress, so as to obtain what we call disorder. In reality we have substituted *will* for the mechanism of nature; we have replaced the "automatic order" by a multitude of elementary wills, just to the extent that we imagine the apparition or vanishing of phenomena. No doubt, for all these little wills to constitute a "willed order," they must have accepted the direction of a higher will. But, on looking closely at them, we see that that is just what they do: our own will is there, which objectifies itself in each of these capricious wills in turn, and takes good care not to connect the same with the same, nor to permit the effect to be proportional to the cause—in fact makes one simple intention hover over the whole of the elementary volitions. Thus, here again, the absence of one of the two orders consists in the presence of the other. In analyzing the idea of chance, which is closely akin to the idea of disorder, we find the same elements. When the wholly mechanical play of the causes which stop the wheel on a number makes me win, and consequently acts like a good genius, careful of my interests, or when the wholly mechanical force of the wind tears a tile off the roof and throws it on to my head, that is to say acts like a bad genius, conspiring against my person: in both cases I find a mechanism where I should have looked for, where, indeed, it seems as if I ought to have found, an intention. That is what I express in speaking of *chance*. And of an anarchical world, in which phenomena succeed each other capriciously, I should say again that it is a realm of chance, meaning that I find before me wills, or rather *decrees*, when what I am expecting is mechanism. Thus is explained the singular vacillation of the mind when it tries to define chance. Neither efficient cause nor final cause can furnish the definition sought. The mind swings to and fro, unable to rest, between the idea of an absence of final cause and that of an absence of efficient cause, each of these definitions sending it back to the other. The problem remains insoluble, in fact, so long as the idea of chance is regarded as a pure idea, without mixture of feeling. But, in reality, chance merely objectifies the state of mind of one who, expecting one of the two kinds of order, finds himself confronted with the other. Chance and disorder are therefore necessarily conceived as relative. So if we wish to represent them to ourselves as absolute, we perceive that we are going to and fro like a shuttle between the two kinds of order, passing into the one just at the moment at which we might catch ourself in the other, and that the supposed absence of all order is really the presence of both, with, besides, the swaying of a mind that cannot rest finally in either. Neither in things nor in our idea of things can there be any question of presenting this disorder as the substratum of order, since it implies the two kinds of order and is made of their combination.

But our intelligence is not stopped by this. By a simple *sic jubeo* it posits a disorder which is an "absence of order." In so doing it thinks a word or a set of words, nothing more. If it seeks to attach an idea to the word, it finds that disorder may indeed be the negation of order, but that this negation is then the implicit affirmation of the presence of the opposite order, which we shut our eyes to because it does not interest us, or which we evade by denying the second order in its turn—that is, at bottom, by re-establishing the first. How can we speak, then, of an incoherent diversity which an understanding organizes? It is no use for us to say that no one supposes this incoherence to be realized or realizable: when we speak of it, we believe we are thinking of it; now, in analyzing the idea actually present, we find, as we said before, only the disappointment of the mind confronted with an order that does not interest it, or a swaying of the mind between two kinds of order, or, finally, the idea pure and simple of the empty word that we have created by joining a negative prefix to a word which itself signifies something. But it is this analysis that we neglect to make. We omit it, precisely because it does not occur to us to distinguish two kinds of order that are irreducible to one another.

We said, indeed, that all order necessarily appears as contingent. If there are two kinds of order, this contingency of order is explained: one of the forms is contingent in relation to the other. Where I find the geometrical order, the vital was possible; where the order is vital, it might have been geometrical. But suppose that the order is everywhere of the same kind, and simply admits of degrees which go from the geometrical to the vital: if a determinate order still appears to me to be contingent, and can no longer be so by relation to an order of another kind, I shall necessarily believe that the order is contingent by relation to an *absence of itself*, that is to say by relation to a state of things "in which there is no order at all." And this state of things I shall believe that I am thinking of, because it is implied, it seems, in the very contingency of order, which is an unquestionable fact. I shall therefore place at the summit of the hierarchy the vital order; then, as a diminution or lower complication of it, the geometrical order; and finally, at the bottom of all, an absence of order, incoherence itself, on which order is superposed. This is why incoherence has the effect on me of a word behind which there must be something real, if not in things, at least in thought. But if I observe that the state of things implied by the contingency of a determinate order is simply the presence of the contrary order, and if by this very fact I posit two kinds of order, each the inverse of the other, I perceive that no intermediate degrees can be imagined between the two orders, and that there is no going down from the two orders to the "incoherent." Either the incoherent is only a word, devoid of meaning, or, if I give it a meaning, it is on condition of putting incoherence midway between the two orders, and not below both of them. There is not first the incoherent, then the geometrical, then the vital; there is only the geometrical and the vital, and then, by a swaying of the mind between them, the idea of the incoherent. To speak of an unco-ordinated diversity to which order is superadded is therefore to commit a veritable *petitio principii*; for in imagining the unco-ordinated we really posit an order, or rather two.

This long analysis was necessary to show how the real can pass from tension to extension and from freedom to mechanical necessity by way of inversion. It was not enough to prove that this relation between the two terms is suggested to us, at once, by consciousness and by sensible experience. It was necessary to prove that the geometrical order has no need of explanation, being purely and simply the suppression of the inverse order. And, for that, it was indispensable to prove that suppression is always a substitution and is even necessarily conceived as such: it is the requirements of practical life alone that suggest to us here a way of speaking that deceives us both as to what happens in things and as to what is present to our thought. We must now examine more closely the inversion whose consequences we have just described. What, then, is the principle that has only to let go its tension—may we say to *detend*—in order to *extend,* the interruption of the cause here being equivalent to a reversal of the effect?

For want of a better word we have called it consciousness. But we do not mean the narrowed consciousness that functions in each of us. Our own consciousness is the consciousness of a certain living being, placed in a certain point of space; and though it does indeed move in the same direction as its principle, it is continually drawn the opposite way, obliged, though it goes forward, to look behind. This retrospective vision is, as we have shown, the natural function of the intellect, and consequently of distinct consciousness. In order that our consciousness shall coincide with something of its principle, it must detach itself from the *already-made* and attach itself to the *being-made*. It needs that, turning back on itself and twisting on itself, the faculty of *seeing* should be made to be one with the act of *willing*—a painful effort which we can make suddenly, doing violence to our nature, but cannot sustain more than a few moments. In free action, when we contract our whole being in order to thrust it forward, we have the more or less clear consciousness of motives and of impelling forces, and even, at rare moments, of the becoming by which they are organized into an act: but the pure willing, the current that runs through this matter, communicating life to it, is a thing which we hardly feel, which at most we brush lightly as it passes. Let us try, however, to install ourselves within it, if only for a moment; even then it is an individual and fragmentary will that we grasp. To get to the principle of all life, as also of all materiality, we must go further still. Is it impossible? No, by no means; the history of philosophy is there to bear witness. There is no durable system that is not, at least in some of its parts, vivified by intuition. Dialectic is necessary to put intuition to the proof, necessary also in order that intuition should break itself up into concepts and so be propagated to other men; but all it does, often enough, is to develop the result of that intuition which transcends it. The truth is, the two procedures are of opposite direction: the same effort, by which ideas are connected with ideas, causes the intuition which the ideas were storing up to vanish. The philosopher is obliged to abandon intuition, once he has received from it the impetus, and to rely on himself to carry on the movement by pushing the concepts one after another. But he soon feels he has lost foothold; he must come into touch with intuition again; he must undo most of what he has done. In short, dialectic is what ensures the agreement of our thought with itself. But by dialectic—which is only a relaxation of intuition—many different agreements are possible, while there is only one truth. Intuition, if it could be prolonged beyond a few instants, would not only make the philosopher agree with his own thought, but also all philosophers with each other. Such as it is, fugitive and incomplete, it is, in each system, what is worth more than the system and survives it. The object of philosophy would be reached if this intuition could be sustained, generalized and, above all, assured of external points of reference in order not to go astray. To that end a continual coming and going is necessary between nature and mind.

When we put back our being into our will, and our will itself into the impulsion it prolongs, we understand, we feel, that reality is a perpetual growth, a creation pursued without end. Our will already performs this miracle. Every human work in which there is invention, every voluntary act in which there is freedom, every movement of an organism that manifests spontaneity, brings something new into the world. True, these are only creations of form. How could they be anything else? We are not the vital current itself; we are this current already loaded with matter, that is, with congealed parts of its own substance which it carries along its course. In the composition of a work of genius, as in a simple free decision, we do, indeed, stretch the spring of our activity to the utmost and thus create what no mere assemblage of materials could have given (what assemblage of curves already known can ever be equivalent to the pencil-stroke of a great artist?) but there are, none the less, elements here that pre-exist and survive their organization. But if a simple arrest of the action that generates form could constitute matter (are not the original lines drawn by the artist themselves already the fixation and, as it were, congealment of a movement?), a creation of matter would be neither incomprehensible nor inadmissible. For we seize from within, we live at every instant, a creation of form, and it is just in those cases in which the form is pure, and in which the creative current is momentarily interrupted, that there is a creation of matter. Consider the letters of the alphabet that enter into the composition of everything that has ever been written: we do not conceive that new letters spring up and come to join themselves to the others in order to make a new poem. But that the poet creates the poem and that human thought is thereby made richer, we understand very well: this creation is a simple act of the mind, and action has only to make a pause, instead of continuing into a new creation, in order that, of itself, it may break up into words which dissociate themselves into letters which are added to all the letters there are already in the world. Thus, that the number of atoms composing the material universe at a given moment should increase runs counter to our habits of mind, contradicts the whole of our experience; but that a reality of quite another order, which contrasts with the atom as the thought of the poet with the letters of the alphabet, should increase by sudden additions, is not inadmissible; and the reverse of each addition might indeed be a world, which we then represent to ourselves, symbolically, as an assemblage of atoms.

The mystery that spreads over the existence of the universe comes in great part from this, that we want the genesis of it to have been accomplished at one stroke or the whole of matter to be eternal. Whether we speak of creation or posit an uncreated matter, it is the totality of the universe that we are considering at once. At the root of this habit of mind lies the prejudice which we will analyze in our next chapter, the idea, common to materialists and to their opponents, that there is no really acting duration, and that the absolute—matter or mind—

can have no place in concrete time, in the time which we feel to be the very stuff of our life. From which it follows that everything is given once for all, and that it is necessary to posit from all eternity either material multiplicity itself, or the act creating this multiplicity, given in block in the divine essence. Once this prejudice is eradicated, the idea of creation becomes more clear, for it is merged in that of growth. But it is no longer then of the universe in its totality that we must speak.

Why should we speak of it? The universe is an assemblage of solar systems which we have every reason to believe analogous to our own. No doubt they are not absolutely independent of one another. Our sun radiates heat and light beyond the farthest planet, and, on the other hand, our entire solar system is moving in a definite direction as if it were drawn. There is, then, a bond between the worlds. But this bond may be regarded as infinitely loose in comparison with the mutual dependence which unites the parts of the same world among themselves; so that it is not artificially, for reasons of mere convenience, that we isolate our solar system: nature itself invites us to isolate it. As living beings, we depend on the planet on which we are, and on the sun that provides for it, but on nothing else. As thinking beings, we may apply the laws of our physics to our own world, and extend them to each of the worlds taken separately; but nothing tells us that they apply to the entire universe, nor even that such an affirmation has any meaning; for the universe is not made, but is being made continually. It is growing, perhaps indefinitely, by the addition of new worlds.

Let us extend, then, to the whole of our solar system

the two most general laws of our science, the principle of conservation of energy and that of its degradation—limiting them, however, to this relatively closed system and to other systems relatively closed. Let us see what will follow. We must remark, first of all, that these two principles have not the same metaphysical scope. The first is a quantitative law, and consequently relative, in part, to our methods of measurement. It says that, in a system presumed to be closed, the total energy, that is to say the sum of its kinetic and potential energy, remains constant. Now, if there were only kinetic energy in the world, or even if there were, besides kinetic energy, only one single kind of potential energy, but no more, the artifice of measurement would not make the law artificial. The law of the conservation of energy would express indeed that *something* is preserved in constant quantity. But there are, in fact, energies of various kinds,[1] and the measurement of each of them has evidently been so chosen as to justify the principle of conservation of energy. Convention, therefore, plays a large part in this principle, although there is undoubtedly, between the variations of the different energies composing one and the same system, a mutual dependence which is just what has made the extension of the principle possible by measurements suitably chosen. If, therefore, the philosopher applies this principle to the solar system complete, he must at least soften its outlines. The law of the conservation of energy cannot here express the objective permanence of a certain quantity of a certain thing, but rather the necessity for every change that is brought about to

[1] On these differences of quality see the work of Duhem, *L'Evolution de la mécanique*, Paris, 1905, pp. 197 ff.

be counterbalanced in some way by a change in an opposite direction. That is to say, even if it governs the whole of our solar system, the law of the conservation of energy is concerned with the relationship of a fragment of this world to another fragment rather than with the nature of the whole.

It is otherwise with the second principle of thermodynamics. The law of the degradation of energy does not bear essentially on magnitudes. No doubt the first idea of it arose, in the thought of Carnot, out of certain quantitative considerations on the yield of thermic machines. Unquestionably, too, the terms in which Clausius generalized it were mathematical, and a calculable magnitude, "entropy," was, in fact, the final conception to which he was led. Such precision is necessary for practical applications. But the law might have been vaguely conceived, and, if absolutely necessary, it might have been roughly formulated, even though no one had ever thought of measuring the different energies of the physical world, even though the concept of energy had not been created. Essentially, it expresses the fact that all physical changes have a tendency to be degraded into heat, and that heat tends to be distributed among bodies in a uniform manner. In this less precise form, it becomes independent of any convention; it is the most metaphysical of the laws of physics since it points out without interposed symbols, without artificial devices of measurements, the direction in which the world is going. It tells us that changes that are visible and heterogeneous will be more and more diluted into changes that are invisible and homogeneous, and that the instability to which we owe the richness and variety of the changes taking place in our solar system will gradually give way to the relative stability of elementary vibrations continually and perpetually repeated. Just so with a man who keeps up his strength as he grows old, but spends it less and less in actions, and comes, in the end, to employ it entirely in making his lungs breathe and his heart beat.

From this point of view, a world like our solar system is seen to be ever exhausting something of the mutability it contains. In the beginning, it had the maximum of possible utilization of energy: this mutability has gone on diminishing unceasingly. Whence does it come? We might at first suppose that it has come from some other point of space, but the difficulty is only set back, and for this external source of mutability the same question springs up. True, it might be added that the number of worlds capable of passing mutability to each other is unlimited, that the sum of mutability contained in the universe is infinite, that there is therefore no ground on which to seek its origin or to foresee its end. A hypothesis of this kind is as irrefutable as it is indemonstrable; but to speak of an infinite universe is to admit a perfect coincidence of matter with abstract space, and consequently an absolute externality of all the parts of matter in relation to one another. We have seen above what we must think of this theory, and how difficult it is to reconcile with the idea of a reciprocal influence of all the parts of matter on one another, an influence to which indeed it itself makes appeal. Again it might be supposed that the general instability has arisen from a general state of stability; that the period in which we now are, and in which the utilizable energy is diminishing, has been preceded by a period in which the mutability was increasing, and that the alternations of increase and diminution succeed each other forever. This hypothesis is theoretically conceivable, as has been demonstrated quite recently; but, according to the calculations of Boltzmann, the mathematical improbability of it passes all imagination and practically amounts to absolute impossibility.[1] In reality, the problem remains insoluble as long as we keep on the ground of physics, for the physicist is obliged to attach energy to extended particles, and, even if he regards the particles only as reservoirs of energy, he remains in space: he would belie his rôle if he sought the origin of these energies in an extra-spatial process. It is there, however, in our opinion, that it must be sought.

Is it extension in general that we are considering *in abstracto*? *Extension*, we said, appears only as a *tension* which is interrupted. Or, are we considering the concrete reality that fills this extension? The order which reigns there, and which is manifested by the laws of nature, is an order which must be born of itself when the inverse order is suppressed; a detension of the will would produce precisely this suppression. Lastly, we find that the direction, which this reality takes, suggests to us the idea of a thing *unmaking itself*; such, no doubt, is one of the essential characters of materiality. What conclusion are we to draw from all this, if not that the process by which this thing *makes itself* is directed in a contrary way to that of physical processes, and that it is therefore, by its very definition, immaterial? The vision we have of the material world is that of a weight which falls: no image drawn from matter, properly so called, will ever give us the idea of the weight rising. But this conclusion will

[1] Boltzmann, *Vorlesungen uber Gastheorie*. Leipzig, 1898, pp. 253 ff.

come home to us with still greater force if we press nearer to the concrete reality, and if we consider, no longer only matter in general, but, within this matter, living bodies.

All our analyses show us, in life, an effort to re-mount the incline that matter descends. In that, they reveal to us the possibility, the necessity even of a process the inverse of materiality, creative of matter by its interruption alone. The life that evolves on the surface of our planet is indeed attached to matter. If it were pure consciousness, *a fortiori* if it were supraconsciousness, it would be pure creative activity. In fact, it is riveted to an organism that subjects it to the general laws of inert matter. But everything happens as if it were doing its utmost to set itself free from these laws. It has not the power to reverse the direction of physical changes, such as the principle of Carnot determines it. It does, however, behave absolutely as a force would behave which, left to itself, would work in the inverse direction. Incapable of *stopping* the course of material changes downwards, it succeeds in *retarding* it. The evolution of life really continues, as we have shown, an initial impulsion: this impulsion, which has determined the development of the chlorophyllian function in the plant and of the sensory-motor system in the animal, brings life to more and more efficient acts by the fabrication and use of more and more powerful explosives. Now, what do these explosives represent if not a storing-up of the solar energy, the degradation of which energy is thus provisionally suspended on some of the points where it was being poured forth? The usable energy which the explosive conceals will be expended, of course, at the moment of the explosion; but it would have been expended sooner if an organism had not happened to be there to arrest its dissipation, in order to retain it and save it up. As we see it today, at the point to which it was brought by a scission of the mutually complementary tendencies which it contained within itself, life is entirely dependent on the chlorophyllian function of the plant. This means that, looked at in its initial impulsion, before any scission, life was a tendency to accumulate in a reservoir, as do especially the green parts of vegetables, with a view to an instantaneous effective discharge, like that which an animal brings about, something that would have otherwise flowed away. It is like an effort to raise the weight which falls. True, it succeeds only in retarding the fall. But at least it can give us an idea of what the raising of the weight was.[1]

Let us imagine a vessel full of steam at a high pressure, and here and there in its sides a crack through which the steam is escaping in a jet. The steam thrown into the air is nearly all condensed into little drops which fall back, and this condensation and this fall represent simply the loss of something, an interruption, a deficit.

[1] In a book rich in facts and in ideas (*La Dissolution opposée à l'évolution*, Paris, 1899), M. André Lalande shows us everything going toward death, in spite of the momentary resistance which organisms seem to oppose.—But, even from the side of unorganized matter, have we the right to extend to the entire universe considerations drawn from the present state of our solar system? Beside the worlds which are dying, there are without doubt worlds that are being born. On the other hand, in the organized world, the death of individuals does not seem at all like a diminution of "life in general," or like a necessity which life submits to reluctantly. As has been more than once remarked, life has never made an effort to prolong indefinitely the existence of the individual, although on so many other points it has made so many successful efforts. Everything is *as if* this death had been willed, or at least accepted, for the greater progress of life in general.

But a small part of the jet of steam subsists, uncondensed, for some seconds; it is making an effort to raise the drops which are falling; it succeeds at most in retarding their fall. So, from an immense reservoir of life, jets must be gushing out unceasingly, of which each, falling back, is a world. The evolution of living species within this world represents what subsists of the primitive direction of the original jet, and of an impulsion which continues itself in a direction the inverse of materiality. But let us not carry too far this comparison. It gives us but a feeble and even deceptive image of reality, for the crack, the jet of steam, the forming of the drops, are determined necessarily, whereas the creation of a world is a free act, and the life within the material world participates in this liberty. Let us think rather of an action like that of raising the arm; then let us suppose that the arm, left to itself, falls back, and yet that there subsists in it, striving to raise it up again, something of the will that animates it. In this image of a *creative action which unmakes itself* we have already a more exact representation of matter. In vital activity we see, then, that which subsists of the direct movement in the inverted movement, *a reality which is making itself in a reality which is unmaking itself.*

Everything is obscure in the idea of creation if we think of *things* which are created and a *thing* which creates, as we habitually do, as the understanding cannot help doing. We shall show the origin of this illusion in our next chapter. It is natural to our intellect, whose function is essentially practical, made to present to us things and states rather than changes and acts. But things and states are only views, taken by our mind, of becoming.

There are no things, there are only actions. More particularly, if I consider the world in which we live, I find that the automatic and strictly determined evolution of this well-knit whole is action which is unmaking itself, and that the unforeseen forms which life cuts out in it, forms capable of being themselves prolonged into unforeseen movements, represent the action that is making itself. Now, I have every reason to believe that the other worlds are analogous to ours, that things happen there in the same way. And I know they were not all constructed at the same time, since observation shows me, even today, nebulae in course of concentration. Now, if the same kind of action is going on everywhere, whether it is that which is unmaking itself or whether it is that which is striving to remake itself, I simply express this probable similitude when I speak of a center from which worlds shoot out like rockets in a fire-works display—provided, however, that I do not present this center as a *thing*, but as a continuity of shooting out. God thus defined, has nothing of the already made; He is unceasing life, action, freedom. Creation, so conceived, is not a mystery; we experience it in ourselves when we act freely. That new things can join things already existing is absurd, no doubt, since the *thing* results from a solidification performed by our understanding, and there are never any things other than those that the understanding has thus constituted. To speak of things creating themselves would therefore amount to saying that the understanding presents to itself more than it presents to itself—a self-contradictory affirmation, an empty and vain idea. But that action increases as it goes on, that it creates in the measure of its advance, is what each of us finds when he watches himself act. Things are constituted by the instantaneous cut which the understanding practices, at a given moment, on a flux of this kind, and what is mysterious when we compare the cuts together becomes clear when we relate them to the flux. Indeed, the modalities of creative action, in so far as it is still going on in the organization of living forms, are much simplified when they are taken in this way. Before the complexity of an organism and the practically infinite multitude of interwoven analyses and syntheses it presupposes, our understanding recoils disconcerted. That the simple play of physical and chemical forces, left to themselves, should have worked this marvel, we find hard to believe. And if it is a profound science which is at work, how are we to understand the influence exercised on this matter without form by this form without matter? But the difficulty arises from this, that we represent statically ready-made material particles juxtaposed to one another, and, also statically, an external cause which plasters upon them a skilfully contrived organization. In reality, life is a movement, materiality is the inverse movement, and each of these two movements is simple, the matter which forms a world being an undivided flux, and undivided also the life that runs through it, cutting out in it living beings all along its track. Of these two currents the second runs counter to the first, but the first obtains, all the same, something from the second. There results between them a *modus vivendi*, which is organization. This organization takes, for our senses and for our intellect, the form of parts entirely external to other parts in space and in time. Not only do we shut our eyes to the unity of the impulse which, passing through generations, links individuals with individuals, species with species, and makes of the whole series of the living one single immense wave flowing over matter, but each individual itself seems to us as an aggregate, aggregate of molecules and aggregate of facts. The reason of this lies in the structure of our intellect, which is formed to act on matter from without, and which succeeds by making, in the flux of the real, instantaneous cuts, each of which becomes, in its fixity, endlessly decomposable. Perceiving, in an organism, only parts external to parts, the understanding has the choice between two systems of explanation only: either to regard the infinitely complex (and thereby infinitely well-contrived) organization as a fortuitous concatenation of atoms, or to relate it to the incomprehensible influence of an external force that has grouped its elements together. But this complexity is the work of the understanding; this incomprehensibility is also its work. Let us try to see, no longer with the eyes of the intellect alone, which grasps only the already made and which looks from the outside, but with the spirit, I mean with that faculty of seeing which is immanent in the faculty of acting and which springs up, somehow, by the twisting of the will on itself, when action is turned into knowledge, like heat, so to say, into light. To movement, then, everything will be restored, and into movement everything will be resolved. Where the understanding, working on the image supposed to be fixed of the progressing action, shows us parts infinitely manifold and an order infinitely well contrived, we catch a glimpse of a simple process, an action which is making itself across an action of the same kind which is unmaking itself, like the fiery path torn by the last rocket of a fire-works display through the black cinders of the spent rockets that are falling dead.

From this point of view, the general considerations we have presented concerning the evolution of life will be cleared up and completed. We will distinguish more sharply what is accidental from what is essential in this evolution.

The impetus of life, of which we are speaking, consists in a need of creation. It cannot create absolutely, because it is confronted with matter, that is to say with the movement that is the inverse of its own. But it seizes upon this matter, which is necessity itself, and strives to introduce into it the largest possible amount of indetermination and liberty. How does it go to work?

An animal high in the scale may be represented in a general way, we said, as a sensory-môtor nervous system imposed on digestive, respiratory, circulatory systems, etc. The function of these latter is to cleanse, repair and protect the nervous system, to make it as independent as possible of external circumstances, but, above all, to furnish it with energy to be expended in movements. The increasing complexity of the organism is therefore due theoretically (in spite of innumerable exceptions due to accidents of evolution) to the necessity of complexity in the nervous system. No doubt, each complication of any part of the organism involves many others in addition, because this part itself must live, and every change in one point of the body reverberates, as it were, throughout. The complication may therefore go on to infinity in all directions; but it is the complication of the nervous system which conditions the others in right, if not always in fact. Now, in what does the progress of the nervous system itself consist? In a simultaneous development of automatic activity and of voluntary activity, the first furnishing the second with an appropriate instrument. Thus, in an organism such as ours, a considerable number of motor mechanisms are set up in the medulla and in the spinal cord, awaiting only a signal to release the corresponding act: the will is employed, in some cases, in setting up the mechanism itself, and in the others in choosing the mechanisms to be released, the manner of combining them and the moment of releasing them. The will of an animal is the more effective and the more intense, the greater the number of the mechanisms it can choose from, the more complicated the switchboard on which all the motor paths cross, or, in other words, the more developed its brain. Thus, the progress of the nervous system assures to the act increasing precision, increasing variety, increasing efficiency and independence. The organism behaves more and more like a machine for action, which reconstructs itself entirely for every new act, as if it were made of india-rubber and could, at any moment, change the shape of all its parts. But, prior to the nervous system, prior even to the organism properly so called, already in the undifferentiated mass of the amoeba, this essential property of animal life is found. The amoeba deforms itself in varying directions; its entire mass does what the differentiation of parts will localize in a sensory-motor system in the developed animal. Doing it only in a rudimentary manner, it is dispensed from the complexity of the higher organisms; there is no need here of the auxiliary elements that pass on to motor elements the energy to expend; the animal moves as a whole, and, as a whole also, procures energy by means of the organic substances it assimilates. Thus, whether low or high in the animal scale, we always find that animal life consists (1) in procuring a provision of energy; (2) in expending it, by means of a matter as supple as possible, in directions variable and unforeseen.

Now, whence comes the energy? From the ingested food, for food is a kind of explosive, which needs only the spark to discharge the energy it stores. Who has made this explosive? The food may be the flesh of an animal nourished on animals and so on; but, in the end it is to the vegetable we always come back. Vegetables alone gather in the solar energy, and the animals do but borrow it from them, either directly or by some passing it on to others. How then has the plant stored up this energy? Chiefly by the chlorophyllian function, a chemicism *sui generis* of which we do not possess the key, and which is probably unlike that of our laboratories. The process consists in using solar energy to fix the carbon of carbonic acid, and thereby to store this energy as we should store that of a water-carrier by employing him to fill an elevated reservoir: the water, once brought up, can set in motion a mill or a turbine, as we will and when we will. Each atom of carbon fixed represents something like the elevation of the weight of water, or like the stretching of an elastic thread uniting the carbon to the oxygen in the carbonic acid. The elastic is relaxed, the weight falls back again, in short the energy held in reserve is restored, when, by a simple release, the carbon is permitted to rejoin its oxygen.

So that all life, animal and vegetable, seems in its essence like an effort to accumulate energy and then to let it flow into flexible channels, changeable in shape, at the end of which it will accomplish infinitely varied kinds of work. That is what the *vital impetus*, passing through matter, would fain do all at once. It would succeed, no doubt, if its power were unlimited, or if some reinforcement could come to it from without. But the impetus is finite, and it has been given once for all. It cannot overcome all obstacles. The movement it starts is sometimes turned aside, sometimes divided, always opposed; and the evolution of the organized world is the unrolling of this conflict. The first great scission that had to be effected was that of the two kingdoms, vegetable and animal, which thus happen to be mutually complementary, without, however, any agreement having been made between them. It is not for the animal that the plant accumulates energy, it is for its own consumption; but its expenditure on itself is less discontinuous, and less concentrated, and therefore less efficacious, than was required by the initial impetus of life, essentially directed toward free actions: the same organism could not with equal force sustain the two functions at once, of gradual storage and sudden use. Of themselves, therefore, and without any external intervention, simply by the effect of the duality of the tendency involved in the original impetus and of the resistance opposed by matter to this impetus, the organisms leaned some in the first direction, others in the second. To this scission there succeeded many others. Hence the diverging lines of evolution, at least what is essential in them. But we must take into account retrogressions, arrests, accidents of every kind. And we must remember, above all, that each species behaves as if the general movement of life stopped at it instead of passing through it. It thinks only of itself, it lives only for itself. Hence the numberless struggles that we behold in nature. Hence a discord, striking and terrible, but for which the original principle of life must not be held responsible.

The part played by contingency in evolution is therefore great. Contingent, generally, are the forms adopted, or rather invented. Contingent, relative to the obstacles encountered in a given place and at a given moment, is the dissociation of the primordial tendency into such and such complementary tendencies which create divergent lines of evolution. Contingent the arrests and setbacks; contingent, in large measure, the adaptations. Two things only are necessary: (1) a gradual accumulation of energy; (2) an elastic canalization of this energy in variable and indeterminable directions, at the end of which are free acts.

This twofold result has been obtained in a particular way on our planet. But it might have been obtained by entirely different means. It was not necessary that life should fix its choice mainly upon the carbon of carbonic acid. What was essential for it was to store solar energy; but, instead of asking the sun to separate, for instance, atoms of oxygen and carbon, it might (theoretically at least, and, apart from practical difficulties possibly insurmountable) have put forth other chemical elements, which would then have had to be associated or dissociated by entirely different physical means. And if the element characteristic of the substances that supply energy to the organism had been other than carbon, the element characteristic of the plastic substances would probably have been other than nitrogen, and the chemistry of living bodies would then have been radically different from what it is. The result would have been living forms without any analogy to those we know, whose anatomy would have been different, whose physiology also would have been different. Alone, the sensory-motor function would have been preserved, if not in its mechanism, at least in its effects. It is therefore probable that life goes on in other planets, in other solar systems also, under forms of which we have no idea, in physical conditions to which it seems to us, from the point of view of our physiology, to be absolutely opposed. If its essential aim is to catch up usable energy in order to expend it in explosive actions, it probably chooses, in each solar system and on each planet, as it does on the earth, the fittest means to get this result in the circumstances with which it is confronted. That is at least what reasoning by analogy leads to, and we use analogy the wrong way when we declare life to be impossible wherever the circumstances with which it is confronted are other than those on the earth. The truth is that life is possible wherever energy descends the incline indicated by Carnot's law and where a cause of inverse direction can retard the descent—that is to say, probably, in all the worlds suspended from all the stars. We go further: it is not even necessary that life should be concentrated and determined in organisms properly so called, that is, in definite bodies presenting to the flow of energy ready-made though elastic canals. It can be conceived (although it can hardly be imagined) that energy might be saved up, and then expended on varying lines running across a matter not yet solidified. Every essential of life would still be there, since there would still be slow accumulation of energy and sudden release. There would hardly be more difference between this vitality, vague and formless, and the definite vitality we know, than there is, in our psychical life, between the state of dream and the state of waking. Such may have been the condition of life in our nebula before the condensation of matter was complete, if it be true that life springs forward at the very moment when, as the effect of an inverse movement, the nebular matter appears.

It is therefore conceivable that life might have assumed a totally different outward appearance and designed forms very different from those we know. With another chemical substratum, in other physical conditions, the impulsion would have remained the same, but it would have split up very differently in course of progress; and the whole would have traveled another road—whether shorter or longer who can tell? In any case, in the entire series of living beings no term would have been what it now is. Now, was it necessary that there should be a series, or terms? Why should not the unique impetus have been impressed on a unique body, which might have gone on evolving?

This question arises, no doubt, from the comparison of life to an impetus. And it must be compared to an impetus, because no image borrowed from the physical world can give more nearly the idea of it. But it is only an image. In reality, life is of the psychological order, and it is of the essence of the psychical to enfold a confused plurality of interpenetrating terms. In space, and in space only, is distinct multiplicity possible: a point is absolutely external to another point. But pure and empty unity, also, is met with only in space; it is that of a mathematical point. Abstract unity and abstract multiplicity are determinations of space or categories of the understanding, whichever we will, spatiality and intellectuality being molded on each other. But what is of psychical nature cannot entirely correspond with space, nor enter perfectly into the categories of the understanding. Is my own person, at a given moment, one or manifold? If I declare it one, inner voices arise and protest—those of the sensations, feelings, ideas, among which my individuality is distributed. But, if I make it distinctly manifold, my consciousness rebels quite as strongly; it affirms that my sensations, my feelings, my thoughts are abstractions which I effect on myself, and that each of my states implies all the others. I am then (we must adopt the language of the understanding, since only the understanding has a language) a unity that is multiple and a multiplicity that is one;[1] but unity and multiplicity are only views of my personality taken by an understanding that directs its categories at me; I enter neither into one nor into the other nor into both at once, although both, united, may give a fair imitation of the mutual interpenetration and continuity that I find at the base of my own self. Such is my inner life, and such also is life in general. While, in its contact with matter, life is comparable to an impulsion or an impetus, regarded in itself it is an immensity of potentiality, a mutual encroachment of thousands and thousands of tendencies which nevertheless are "thousands and thousands" only when once regarded as outside of each other, that is, when spatialized. Contact with matter is what determines this dissociation.

[1] We have dwelt on this point in an article entitled "Introduction à la métaphysique" (*Revue de métaphysique et de morale*, January 1903, pp. 1-25).

Matter divides actually what was but potentially manifold; and, in this sense, individuation is in part the work of matter, in part the result of life's own inclination. Thus, a poetic sentiment, which bursts into distinct verses, lines and words, may be said to have already contained this multiplicity of individuated elements, and yet, in fact, it is the materiality of language that creates it.

But through the words, lines and verses runs the simple inspiration which is the whole poem. So, among the dissociated individuals, one life goes on moving: everywhere the tendency to individualize is opposed and at the same time completed by an antagonistic and complementary tendency to associate, as if the manifold unity of life, drawn in the direction of multiplicity, made so much the more effort to withdraw itself on to itself. A part is no sooner detached than it tends to reunite itself, if not to all the rest, at least to what is nearest to it. Hence, throughout the whole realm of life, a balancing between individuation and association. Individuals join together into a society; but the society, as soon as formed, tends to melt the associated individuals into a new organism, so as to become itself an individual, able in its turn to be part and parcel of a new association. At the lowest degree of the scale of organisms we already find veritable associations, microbial colonies, and in these associations, according to a recent work, a tendency to individuate by the constitution of a nucleus.[1] The same tendency is met with again at a higher stage, in the protophytes, which, once having quitted the parent cell

[1] Cf. a paper written (in Russian) by Serkovski, and reviewed in the *Année biologique*, 1898, p. 317.

by way of division, remain united to each other by the gelatinous substance that surrounds them—also in those protozoa which begin by mingling their pseudopodia and end by welding themselves together. The "colonial" theory of the genesis of higher organisms is well known. The protozoa, consisting of one single cell, are supposed to have formed, by assemblage, aggregates which, relating themselves together in their turn, have given rise to aggregates of aggregates; so organisms more and more complicated, and also more and more differentiated, are born of the association of organisms barely differentiated and elementary.[1] In this extreme form, the theory is open to grave objections: more and more the idea seems to be gaining ground, that polyzoism is an exceptional and abnormal fact.[2] But it is none the less true that things happen *as if* every higher organism was born of an association of cells that have subdivided the work between them. Very probably it is not the cells that have made the individual by means of association; it is rather the individual that has made the cells by means of dissociation.[3] But this itself reveals to us, in the genesis of the individual, a haunting of the social form, as if the individual could develop only on the condition that its substance should be split up into elements having themselves an appearance of individuality and united among themselves by an appearance of sociality. There are numerous

[1] Ed. Perrier, *Les Colonies animales*, Paris, 1897 (2nd edition).

[2] Delage, *L'Hérédité*, 2nd edition, Paris, 1903, p. 97. Cf. by the same author, "La Conception polyzoïque des êtres" (*Revue scientifique*, 1896, pp. 641-653).

[3] This is the theory maintained by Kunstler, Delage, Sedgwick, Labbé, etc. Its development, with bibliographical references, will be found in the work of Busquet, *Les êtres vivants*, Paris, 1899.

cases in which nature seems to hesitate between the two forms, and to ask herself if she shall make a society or an individual. The slightest push is enough, then, to make the balance weigh on one side or the other. If we take an infusorian sufficiently large, such as the Stentor, and cut it into two halves each containing a part of the nucleus, each of the two halves will generate an independent Stentor; but if we divide it incompletely, so that a protoplasmic communication is left between the two halves, we shall see them execute, each from its side, corresponding movements: so that in this case it is enough that a thread should be maintained or cut in order that life should affect the social or the individual form. Thus, in rudimentary organisms consisting of a single cell, we already find that the apparent individuality of the whole is the composition of an *undefined* number of potential individualities potentially associated. But, from top to bottom of the series of living beings, the same law is manifested. And it is this that we express when we say that unity and multiplicity are categories of inert matter, that the vital impetus is neither pure unity nor pure multiplicity, and that if the matter to which it communicates itself compels it to choose one of the two, its choice will never be definitive: it will leap from one to the other indefinitely. The evolution of life in the double direction of individuality and association has therefore nothing accidental about it: it is due to the very nature of life.

Essential also is the progress to reflection. If our analysis is correct, it is consciousness, or rather supra-consciousness, that is at the origin of life. Consciousness, or supra-consciousness, is the name for the rocket whose extinguished fragments fall back as matter; consciousness, again, is the name for that which subsists of the rocket itself, passing through the fragments and lighting them up into organisms. But this consciousness, which is a *need of creation*, is made manifest to itself only where creation is possible. It lies dormant when life is condemned to automatism; it wakens as soon as the possibility of a choice is restored. That is why, in organisms unprovided with a nervous system, it varies according to the power of locomotion and of deformation of which the organism disposes. And in animals with a nervous system, it is proportional to the complexity of the switchboard on which the paths called sensory and the paths called motor intersect—that is, of the brain. How must this solidarity between the organism and consciousness be understood?

We will not dwell here on a point that we have dealt with in former works. Let us merely recall that a theory such as that according to which consciousness is attached to certain neurons, and is thrown off from their work like a phosphorescence, may be accepted by the scientist for the detail of analysis; it is a convenient mode of expression. But it is nothing else. In reality, a living being is a center of action. It represents a certain sum of contingency entering into the world, that is to say, a certain quantity of possible action—a quantity variable with individuals and especially with species. The nervous system of an animal marks out the flexible lines on which its action will run (although the potential energy is accumulated in the muscles rather than in the nervous system itself); its nervous centers indicate, by their development and their configuration, the more or less extended choice it will have among more or less numerous and complicated actions. Now, since the awakening of consciousness in a living creature is the more complete, the greater the latitude of choice allowed to it and the larger the amount of action bestowed upon it, it is clear that the development of consciousness will appear to be dependent on that of the nervous centers. On the other hand, every state of consciousness being, in one aspect of it, a question put to the motor activity and even the beginning of a reply, there is no psychical event that does not imply the entry into play of the cortical mechanisms. Everything seems, therefore, to happen *as if* consciousness sprang from the brain, and *as if* the detail of conscious activity were modeled on that of the cerebral activity. In reality, consciousness does not spring from the brain; but brain and consciousness correspond because equally they measure, the one by the complexity of its structure and the other by the intensity of its awareness, the quantity of *choice* that the living being has at its disposal.

It is precisely because a cerebral state expresses simply what there is of nascent action in the corresponding psychical state, that the psychical state tells us more than the cerebral state. The consciousness of a living being, as we have tried to prove elsewhere, is inseparable from its brain in the sense in which a sharp knife is inseparable from its edge: the brain is the sharp edge by which consciousness cuts into the compact tissue of events, but the brain is no more coextensive with consciousness than the edge is with the knife. Thus, from the fact that two brains, like that of the ape and that of the man, are verv much alike, we cannot conclude that the corresponding consciousnesses are comparable or commensurable.

But the two brains may perhaps be less alike than we suppose. How can we help being struck by the fact that, while man is capable of learning any sort of exercise, of constructing any sort of object, in short of acquiring any kind of motor habit whatsoever, the faculty of combining new movements is strictly limited in the best-endowed animal, even in the ape? The cerebral characteristic of man is there. The human brain is made, like every brain, to set up motor mechanisms and to enable us to choose among them, at any instant, the one we shall put in motion by the pull of a trigger. But it differs from other brains in this, that the number of mechanisms it can set up, and consequently the choice that it gives as to which among them shall be released, is unlimited. Now, from the limited to the unlimited there is all the distance between the closed and the open. It is not a difference of degree, but of kind.

Radical therefore, also, is the difference between animal consciousness, even the most intelligent, and human consciousness. For consciousness corresponds exactly to the living being's power of choice; it is coextensive with the fringe of possible action that surrounds the real action: consciousness is synonymous with invention and with freedom. Now, in the animal, invention is never anything but a variation on the theme of routine. Shut up in the habits of the species, it succeeds, no doubt, in enlarging them by its individual initiative; but it escapes automatism only for an instant, for just the time to create a new automatism. The gates of its prison close as soon as they are opened; by pulling at its chain it succeeds only in stretching it. With man, consciousness breaks the chain. In man, and in man alone, it sets itself free. The whole history of life until man has been that of the effort of consciousness to raise matter, and of the more or less complete overwhelming of consciousness by the matter which has fallen back on it. The enterprise was paradoxical, if, indeed, we may speak here otherwise than by metaphor of enterprise and of effort. It was to create with matter, which is necessity itself, an instrument of freedom, to make a machine which should triumph over mechanism, and to use the determinism of nature to pass through the meshes of the net which this very determinism had spread. But, everywhere except in man, consciousness has let itself be caught in the net whose meshes it tried to pass through: it has remained the captive of the mechanisms it has set up. Automatism, which it tries to draw in the direction of freedom, winds about it and drags it down. It has not the power to escape, because the energy it has provided for acts is almost all employed in maintaining the infinitely subtle and essentially unstable equilibrium into which it has brought matter. But man not only maintains his machine, he succeeds in using it as he pleases. Doubtless he owes this to the superiority of his brain, which enables him to build an unlimited number of motor mechanisms, to oppose new habits to the old ones unceasingly, and, by dividing automatism against itself, to rule it. He owes it to his language, which furnishes consciousness with an immaterial body in which to incarnate itself and thus exempts it from dwelling exclusively on material bodies, whose flux would soon drag it along and finally swallow it up. He owes it to social life, which stores and preserves efforts as language stores thought, fixes thereby a mean level to which individuals must raise themselves at the outset, and by this initial stimulation prevents the average man from slumbering and drives the superior man to mount still higher. But our brain, our society, and our language are only the external and various signs of one and the same internal superiority. They tell, each after its manner, the unique, exceptional success which life has won at a given moment of its evolution. They express the difference of kind, and not only of degree, which separates man from the rest of the animal world. They let us guess that, while at the end of the vast spring-board from which life has taken its leap, all the others have stepped down, finding the cord stretched too high, man alone has cleared the obstacle.

It is in this quite special sense that man is the "term" and the "end" of evolution. Life, we have said, transcends finality as it transcends the other categories. It is essentially a current sent through matter, drawing from it what it can. There has not, therefore, properly speaking, been any project or plan. On the other hand, it is abundantly evident that the rest of nature is not for the sake of man: we struggle like the other species, we have struggled against other species. Moreover, if the evolution of life had encountered other accidents in its course, if, thereby, the current of life had been otherwise divided, we should have been, physically and morally, far different from what we are. For these various reasons it would be wrong to regard humanity, such as we have it before our eyes, as pre-figured in the evolutionary movement. It cannot even be said to be the outcome of the whole of evolution, for evolution has been accomplished on several divergent lines, and while the human species is at the end of one of them, other lines have been followed with other species at their end. It is in a quite different sense that we hold humanity to be the ground of evolution.

From our point of view, life appears in its entirety as an immense wave which, starting from a center, spreads outwards, and which on almost the whole of its circumference is stopped and converted into oscillation: at one single point the obstacle has been forced, the impulsion has passed freely. It is this freedom that the human form registers. Everywhere but in man, consciousness has had to come to a stand; in man alone it has kept on its way. Man, then, continues the vital movement indefinitely, although he does not draw along with him all that life carries in itself. On other lines of evolution there have traveled other tendencies which life implied, and of which, since everything interpenetrates, man has, doubtless, kept something, but of which he has kept only very little. *It is as if a vague and formless being, whom we may call, as we will,* man *or* superman, *had sought to realize himself, and had succeeded only by abandoning a part of himself on the way.* The losses are represented by the rest of the animal world, and even by the vegetable world, at least in what these have that is positive and above the accidents of evolution.

From this point of view, the discordances of which nature offers us the spectacle are singularly weakened. The organized world as a whole becomes as the soil on which was to grow either man himself or a being who morally must resemble him. The animals, however distant they may be from our species, however hostile to it, have none the less been useful traveling companions, on whom consciousness has unloaded whatever encumbrances it was dragging along, and who have enabled it to rise, in man, to heights from which it sees an unlimited horizon open again before it.

It is true that it has not only abandoned cumbersome baggage on the way; it has also had to give up valuable goods. Consciousness, in man, is pre-eminently intellect. It might have been, it ought, so it seems, to have been also intuition. Intuition and intellect represent two opposite directions of the work of consciousness: intuition goes in the very direction of life, intellect goes in the inverse direction, and thus finds itself naturally in accordance with the movement of matter. A complete and perfect humanity would be that in which these two forms of conscious activity should attain their full development. And, between this humanity and ours, we may conceive any number of possible stages, corresponding to all the degrees imaginable of intelligence and of intuition. In this lies the part of contingency in the mental structure of our species. A different evolution might have led to a humanity either more intellectual still or more intuitive. In the humanity of which we are a part, intuition is, in fact, almost completely sacrificed to intellect. It seems that to conquer matter, and to reconquer its own self, consciousness has had to exhaust the best part of its power. This conquest, in the particular conditions in which it has been accomplished, has required that consciousness should adapt itself to the habits of matter and concentrate all its attention on them, in fact determine itself more especially as intellect. Intuition is there, however, but vague and above all discontinuous. It is a lamp almost extinguished, which only glimmers now and then, for a few moments at most. But it glimmers wherever a vital interest is at stake. On our personality, on our liberty, on the place we occupy in the whole of nature, on our origin and perhaps also on our destiny, it throws a light feeble and vacillating, but which none the less pierces the darkness of the night in which the intellect leaves us.

These fleeting intuitions, which light up their object only at distant intervals, philosophy ought to seize, first to sustain them, then to expand them and so unite them together. The more it advances in this work, the more will it perceive that intuition is mind itself, and, in a certain sense, life itself: the intellect has been cut out of it by a process resembling that which has generated matter. Thus is revealed the unity of the spiritual life. We recognize it only when we place ourselves in intuition in order to go from intuition to the intellect, for from the intellect we shall never pass to intuition.

Philosophy introduces us thus into the spiritual life. And it shows us at the same time the relation of the life of the spirit to that of the body. The great error of the doctrines on the spirit has been the idea that by isolating the spiritual life from all the rest, by suspending it in space as high as possible above the earth, they were placing it beyond attack, as if they were not thereby simply exposing it to be taken as an effect of mirage! Certainly they are right to listen to conscience when conscience affirms human freedom; but the intellect is there, which says that the cause determines its effect, that like conditions like, that all is repeated and that all is given. They are right to believe in the absolute reality of the person and in his independence toward matter; but science is there, which shows the interdependence of conscious life and cerebral activity. They are right to attribute to man a privileged place in nature, to hold that the distance is infinite between the animal and man; but the history of life is there, which makes us witness the genesis of species by gradual transformation, and seems thus to reintegrate man in animality. When a strong instinct assures the probability of personal survival, they are right not to close their ears to its voice; but if there exist "souls" capable of an independent life, whence do they come? When, how and why do they enter into this body which we see arise, quite naturally, from a mixed cell derived from the bodies of its two parents? All these questions will remain unanswered, a philosophy of intuition will be a negation of science, will be sooner or later swept away by science, if it does not resolve to see the life of the body just where it really is, on the road that leads to the life of the spirit. But it will then no longer have to do with definite living beings. Life as a whole, from the initial impulsion that thrust it into the world, will appear as a wave which rises, and which is opposed by the descending movement of matter. On the greater part of its surface, at different heights, the current is converted by matter into a vortex. At one point alone it passes freely, dragging with it the obstacle which will weigh on its progress but will not stop it. At this point is humanity; it is our privileged situation. On the other hand, this rising wave is consciousness, and, like all consciousness, it includes potentialities without number which interpenetrate and to which consequently neither the category of unity nor that of multiplicity is appropriate, made as they both are for inert matter. The matter that it bears along with it, and in the interstices of which it inserts itself, alone can divide it into distinct individualities. On flows the current, running through human generations, subdividing itself into individuals. This subdivision was vaguely indicated in it, but could not have been made clear without matter. Thus souls are continually being created, which, nevertheless, in a certain sense pre-existed. They are nothing else than the little rills into which the great river of life divides itself, flowing through the body of humanity. The movement of the stream is distinct from the river bed, although it must adopt its winding course. Consciousness is distinct from the organism it animates, although it must undergo its vicissitudes. As the possible actions which a state of consciousness indicates are at every instant beginning to be carried out in the nervous centers, the brain underlies at every instant the motor indications of the state of consciousness; but the interdependency of consciousness and brain is limited to this; the destiny of consciousness is not bound up on that account with the destiny of cerebral matter. Finally, consciousness is essentially free; it is freedom itself; but it cannot pass through matter without settling on it, without adapting itself to it: this adaptation is what we call intellectuality; and the intellect, turning itself back toward active, that is to say free, consciousness, naturally makes it enter into the conceptual forms into which it is accustomed to see matter fit. It will therefore always perceive freedom in the form of necessity; it will always neglect the part of novelty or of creation inherent in the free act; it will always substitute for action itself an imitation artificial, approximative, obtained by compounding the old with the old and the same with the same. Thus, to the eyes of a philosophy that attempts to reabsorb intellect in intuition, many difficulties vanish or become light. But such a doctrine does not only facilitate speculation; it gives us also more power to act and to live. For, with it, we feel ourselves no longer isolated in humanity, humanity no longer seems isolated in the nature that it dominates. As the smallest grain of dust is bound up with our entire solar system, drawn along with it in that undivided movement of descent which is materiality itself, so all organized beings, from the humblest to the highest, from the first origins of life to the time in which we are, and in all places as in all times, do but evidence a single impulsion, the inverse of the movement of matter, and in itself indivisible. All the living hold together, and all yield to the same tremendous push. The animal takes its stand on the plant, man bestrides animality, and the whole of humanity, in space and in time, is one immense army galloping beside and before and behind each of us in an overwhelming charge able to beat down every resistance and clear the most formidable obstacles, perhaps even death.

CHAPTER IV

THE CINEMATOGRAPHICAL MECHANISM OF THOUGHT AND THE MECHANISTIC ILLUSION—A GLANCE AT THE HISTORY OF SYSTEMS[1]—REAL BECOMING AND FALSE EVOLUTIONISM.

It remains for us to examine in themselves two theoretical illusions which we have frequently met with before, but whose consequences rather than principle have hitherto concerned us. Such is the object of the present chapter. It will afford us the opportunity of removing certain objections, of clearing up certain misunderstandings, and, above all, of defining more precisely, by contrasting it with others, a philosophy which sees in duration the very stuff of reality.

Matter or mind, reality has appeared to us as a perpetual becoming. It makes itself or it unmakes itself, but it is never something made. Such is the intuition that we have of mind when we draw aside the veil which is interposed between our consciousness and ourselves. This, also, is what our intellect and senses themselves would

[1] The part of this chapter which treats of the history of systems, particularly of the Greek philosophy, is only the very succinct résumé of views that we developed at length, from 1900 to 1904, in our lectures at the Collège de France, especially in a course on the *History of the Idea of Time* (1902-1903). We then compared the mechanism of conceptual thought to that of the cinematograph. We believe the comparison will be useful here.

show us of matter, if they could obtain a direct and disinterested idea of it. But, preoccupied before everything with the necessities of action, the intellect, like the senses, is limited to taking, at intervals, views that are instantaneous and by that very fact immobile of the becoming of matter. Consciousness, being in its turn formed on the intellect, sees clearly of the inner life what is already made, and only feels confusedly the making. Thus, we pluck out of duration those moments that interest us, and that we have gathered along its course. These alone we retain. And we are right in so doing, while action only is in question. But when, in *speculating* on the *nature* of the real, we go on regarding it as our practical interest requires us to regard it, we become unable to perceive the true evolution, the radical becoming. Of becoming we perceive only states, of duration only instants, and even when we speak of duration and of becoming, it is of another thing that we are thinking. Such is the most striking of the two illusions we wish to examine. It consists in supposing that we can think the unstable by means of the stable, the moving by means of the immobile.

The other illusion is near akin to the first. It has the same origin, being also due to the fact that we import into speculation a procedure made for practice. All action aims at getting something that we feel the want of, or at creating something that does not yet exist. In this very special sense, it fills a void, and goes from the empty to the full, from an absence to a presence, from the unreal to the real. Now the unreality which is here in question is purely relative to the direction in which our attention is engaged, for we are immersed in realities and cannot pass out of them; only, if the present reality is not the one we are seeking, we speak of the *absence* of this sought-for reality wherever we find the *presence* of another. We thus express what we have as a function of what we want. This is quite legitimate in the sphere of action. But, whether we will or no, we keep to this way of speaking, and also of thinking, when we speculate on the nature of things independently of the interest they have for us. Thus arises the second of the two illusions. We propose to examine this first. It is due, like the other, to the static habits that our intellect contracts when it prepares our action on things. Just as we pass through the immobile to go to the moving, so we make use of the void in order to think the full.

We have met with this illusion already in dealing with the fundamental problem of knowledge. The question, we then said, is to know why there is order, and not disorder, in things. But the question has meaning only if we suppose that disorder, understood as an absence of order, is possible, or imaginable, or conceivable. Now, it is only order that is real; but, as order can take two forms, and as the presence of the one may be said to consist in the absence of the other, we speak of disorder whenever we have before us that one of the two orders for which we are not looking. The idea of disorder is then entirely practical. It corresponds to the disappointment of a certain expectation, and it does not denote the absence of all order, but only the presence of that order which does not offer us actual interest. So that whenever we try to deny order completely, absolutely, we find that we are leaping from one kind of order to the other indefinitely, and that the supposed suppression of the one and the other implies the presence of the two. Indeed, if we go on, and persist in shutting our eyes to this movement of the mind and all it involves, we are no longer dealing with an idea; all that is left of disorder is a word. Thus the problem of knowledge is complicated, and possibly made insoluble, by the idea that order fills a void and that its actual presence is superposed on its virtual absence. We go from absence to presence, from the void to the full, in virtue of the fundamental illusion of our understanding. That is the error of which we noticed one consequence in our last chapter. As we then anticipated, we must come to close quarters with this error, and finally grapple with it. We must face it in itself, in the radically false conception which it implies of negation, of the void and of the nought.[1]

Philosophers have paid little attention to the idea of the nought. And yet it is often the hidden spring, the invisible mover of philosophical thinking. From the first awakening of reflection, it is this that pushes to the fore, right under the eyes of consciousness, the torturing problems, the questions that we cannot gaze at without feeling giddy and bewildered. I have no sooner commenced to philosophize than I ask myself why I exist; and when I take account of the intimate connection in which I stand to the rest of the universe, the difficulty is only pushed back, for I want to know why the universe exists; and if I refer the universe to a Principle immanent or transcendent that supports it or creates it, my thought rests on this principle only a few moments, for the same problem recurs, this time in its full breadth and gener-

[1] The analysis of the idea of the nought which we give here (pp. 299-324) has appeared before in the *Revue philosophique* (November 1906).

ality: Whence comes it, and how can it be understood, that anything exists? Even here, in the present work, when matter has been defined as a kind of descent, this descent as the interruption of a rise, this rise itself as a growth, when finally a Principle of creation has been put at the base of things, the same question springs up: How —why does this principle exist rather than nothing?

Now, if I push these questions aside and go straight to what hides behind them, this is what I find:—Existence appears to me like a conquest over nought. I say to myself that there might be, that indeed there ought to be, nothing, and I then wonder that there is something. Or I represent all reality extended on nothing as on a carpet: at first was nothing, and being has come by superaddition to it. Or, yet again, if something has always existed, nothing must always have served as its substratum or receptacle, and is therefore eternally prior. A glass may have always been full, but the liquid it contains nevertheless fills a void. In the same way, being may have always been there, but the nought which is filled, and, as it were, stopped up by it, pre-exists for it none the less, if not in fact at least in right. In short, I cannot get rid of the idea that the full is an embroidery on the canvas of the void, that being is superimposed on nothing, and that in the idea of "nothing" there is *less* than in that of "something." Hence all the mystery.

It is necessary that this mystery should be cleared up. It is more especially necessary, if we put duration and free choice at the base of things. For the disdain of metaphysics for all reality that endures comes precisely from this, that it reaches being only by passing through "not-being," and that an existence which endures seems to it not strong enough to conquer non-existence and itself posit itself. It is for this reason especially that it is inclined to endow true being with a *logical,* and not a psychological or a physical existence. For the nature of a purely logical existence is such that it seems to be self-sufficient and to posit itself by the effect alone of the force immanent in truth. If I ask myself why bodies or minds exist rather than nothing, I find no answer; but that a logical principle, such as A = A, should have the power of creating itself, triumphing over the nought throughout eternity, seems to me natural. A circle drawn with chalk on a blackboard is a thing which needs explanation: this entirely physical existence has not by itself wherewith to vanquish non-existence. But the "logical essence" of the circle, that is to say, the possibility of drawing it according to a certain law—in short, its definition—is a thing which appears to me eternal: it has neither place nor date; for nowhere, at no moment, has the drawing of a circle begun to be possible. Suppose, then, that the principle on which all things rest, and which all things manifest possesses an existence of the same nature as that of the definition of the circle, or as that of the axiom A = A: the mystery of existence vanishes, for the being that is at the base of everything posits itself then in eternity, as logic itself does. True, it will cost us rather a heavy sacrifice: if the principle of all things exists after the manner of a logical axiom or of a mathematical definition, the things themselves must go forth from this principle like the applications of an axiom or the consequences of a definition, and there will no longer be place, either in the things or in their principle, for efficient causality understood in the sense of a free choice. Such are precisely the conclusions of a doctrine like that of Spinoza, or even that of Leibniz, and such indeed has been their genesis.

Now, if we could prove that the idea of the nought, in the sense in which we take it when we oppose it to that of existence, is a pseudo-idea, the problems that are raised around it would become pseudo-problems. The hypothesis of an absolute that acts freely, that in an eminent sense endures, would no longer raise up intellectual prejudices. The road would be cleared for a philosophy more nearly approaching intuition, and which would no longer ask the same sacrifices of common sense.

Let us then see what we are thinking about when we speak of "Nothing." To represent "Nothing," we must either imagine it or conceive it. Let us examine what this image or this idea may be. First, the image.

I am going to close my eyes, stop my ears, extinguish one by one the sensations that come to me from the outer world. Now it is done; all my perceptions vanish, the material universe sinks into silence and the night.—I subsist, however, and cannot help myself subsisting. I am still there, with the organic sensations which come to me from the surface and from the interior of my body, with the recollections which my past perceptions have left behind them—nay, with the impression, most positive and full, of the void I have just made about me. How can I suppress all this? How eliminate myself? I can even, it may be, blot out and forget my recollections up to my immediate past; but at least I keep the consciousness of my present reduced to its extremest poverty, that is to say, of the actual state of my body. I will try, however, to do away even with this consciousness itself. I will reduce more and more the sensations my body sends in to me: now they are almost gone; now they are gone, they have disappeared in the night where all things else have already died away. But no! At the very instant that my consciousness is extinguished, another consciousness lights up—or rather, it was already alight: it had arisen the instant before, in order to witness the extinction of the first; for the first could disappear only for another and in the presence of another. I see myself annihilated only if I have already resuscitated myself by an act which is positive, however involuntary and unconscious. So, do what I will, I am always perceiving something, either from without or from within. When I no longer know anything of external objects, it is because I have taken refuge in the consciousness that I have of myself. If I abolish this inner self, its very abolition becomes an object for an imaginary self which now perceives as an external object the self that is dying away. Be it external or internal, some object there always is that my imagination is representing. My imagination, it is true, can go from one to the other, I can by turns imagine a nought of external perception or a nought of internal perception, but not both at once, for the absence of one consists, at bottom, in the exclusive presence of the other. But, from the fact that two relative noughts are imaginable in turn, we wrongly conclude that they are imaginable together: a conclusion the absurdity of which must be obvious, for we cannot imagine a nought without perceiving, at least confusedly, that we are imagining it, consequently that we are acting, that we are thinking, and therefore that something still subsists.

The image, then. properly so called, of a suppression

of everything is never formed by thought. The effort by which we strive to create this image simply ends in making us swing to and fro between the vision of an outer and that of an inner reality. In this coming and going of our mind between the without and the within, there is a point, at equal distance from both, in which it seems to us that we no longer perceive the one, and that we do not yet perceive the other: it is there that the image of "Nothing" is formed. In reality, we then perceive both, having reached the point where the two terms come together, and the image of Nothing, so defined, is an image full of things, an image that includes at once that of the subject and that of the object and, besides, a perpetual leaping from one to the other and the refusal ever to come to rest finally on either. Evidently this is not the nothing that we can oppose to being, and put before or beneath being, for it already includes existence in general.

But we shall be told that, if the representation of Nothing, visible or latent, enters into the reasonings of philosophers, it is not as an image, but as an idea. It may be agreed that we do not imagine the annihilation of everything, but it will be claimed that we can conceive it. We conceive a polygon with a thousand sides, said Descartes, although we do not see it in imagination: it is enough that we can clearly represent the possibility of constructing it. So with the idea of the annihilation of everything. Nothing simpler, it will be said, than the procedure by which we construct the idea of it. There is, in fact, not a single object of our experience that we cannot suppose annihilated. Extend this annihilation of a first object to a second, then to a third, and so on as long as you please: the nought is the limit toward which the operation tends. And the nought so defined is the annihilation of everything. That is the theory. We need only consider it in this form to see the absurdity it involves.

An idea constructed by the mind is an idea only if its pieces are capable of coexisting; it is reduced to a mere word if the elements that we bring together to compose it are driven away as fast as we assemble them. When I have defined the circle, I easily represent a black or a white circle, a circle in cardboard, iron, or brass, a transparent or an opaque circle—but not a square circle, because the law of the generation of the circle excludes the possibility of defining this figure with straight lines. So my mind can represent any existing thing whatever as annihilated;—but if the annihilation of anything by the mind is an operation whose mechanism implies that it works on a part of the whole, and not on the whole itself, then the extension of such an operation to the totality of things becomes self-contradictory and absurd, and the idea of an annihilation of everything presents the same character as that of a square circle: it is not an idea, it is only a word. So let us examine more closely the mechanism of the operation.

In fact, the object suppressed is either external or internal: it is a thing or it is a state of consciousness. Let us consider the first case. I annihilate in thought an external object: in the place where it was, there is no longer anything.—No longer anything of that object, of course, but another object has taken its place: there is no absolute void in nature. But admit that an absolute void is possible: it is not of that void that I am thinking when I say that the object, once annihilated. leaves its place unoccupied; for by the hypothesis it is a *place*, that is a void limited by precise outlines, or, in other words, a kind of *thing*. The void of which I speak, therefore, is, at bottom, only the absence of some definite object, which was here at first, is now elsewhere and, in so far as it is no longer in its former place, leaves behind it, so to speak, the void of itself. A being unendowed with memory or prevision would not use the words "void" or "nought"; he would express only what is and what is perceived; now, what is, and what is perceived, is the *presence* of one thing or of another, never the *absence* of anything. There is absence only for a being capable of remembering and expecting. He remembered an object, and perhaps expected to encounter it again; he finds another, and he expresses the disappointment of his expectation (an expectation sprung from recollection) by saying that he no longer finds anything, that he encounters "nothing." Even if he did not expect to encounter the object, it is a possible expectation of it, it is still the falsification of his eventual expectation that he expresses by saying that the object is no longer where it was. What he perceives in reality, what he will succeed in effectively thinking of, is the presence of the old object in a new place or that of a new object in the old place; the rest, all that is expressed negatively by such words as "nought" or the "void," is not so much thought as feeling, or, to speak more exactly, it is the tinge that feeling gives to thought. The idea of annihilation or of partial nothingness is therefore formed here in the course of the substitution of one thing for another, whenever this substitution is thought by a mind that would prefer to keep the old thing in the place of the new, or at least conceives this preference as possible. The idea implies on the subjective side a preference, on the objective side a substitution, and is nothing else but a combination of, or rather an interference between, this feeling of preference and this idea of substitution.

Such is the mechanism of the operation by which our mind annihilates an object and succeeds in representing in the external world a partial nought. Let us now see how it represents it within itself. We find in ourselves phenomena that are produced, and not phenomena that are not produced. I experience a sensation or an emotion, I conceive an idea, I form a resolution: my consciousness perceives these facts, which are so many *presences*, and there is no moment in which facts of this kind are not present to me. I can, no doubt, interrupt by thought the course of my inner life; I may suppose that I sleep without dreaming or that I have ceased to exist; but at the very instant when I make this supposition, I conceive myself, I imagine myself watching over my slumber or surviving my annihilation, and I give up perceiving myself from within only by taking refuge in the perception of myself from without. That is to say that here again the full always succeeds the full, and that an intelligence that was only intelligence, that had neither regret nor desire, whose movement was governed by the movement of its object, could not even conceive an absence or a void. The conception of a void arises here when consciousness, lagging behind itself, remains attached to the recollection of an old state when another state is already present. It is only a comparison between what is and what could or ought to be, between the full and the full. In a word, whether it be a void of matter or a void of consciousness, *the representation of the void is always a representation which is full and which resolves itself on analysis into two positive elements: the idea, distinct or confused, of a substitution, and the feeling, experienced or imagined, of a desire or a regret.*

It follows from this double analysis that the idea of the absolute nought, in the sense of the annihilation of everything, is a self-destructive idea, a pseudo-idea, a mere word. If suppressing a thing consists in replacing it by another, if thinking the absence of one thing is only possible by the more or less explicit representation of the presence of some other thing, if, in short, annihilation signifies before anything else substitution, the idea of an "annihilation of everything" is as absurd as that of a square circle. The absurdity is not obvious, because there exists no particular object that cannot be supposed annihilated; then, from the fact that there is nothing to prevent each thing in turn being suppressed in thought, we conclude that it is possible to suppose them suppressed altogether. We do not see that suppressing each thing in turn consists precisely in replacing it in proportion and degree by another, and therefore that the suppression of absolutely everything implies a downright contradiction in terms, since the operation consists in destroying the very condition that makes the operation possible.

But the illusion is tenacious. Though suppressing one thing consists *in fact* in substituting another for it, we do not conclude, we are unwilling to conclude, that the annihilation of a thing *in thought* implies the substitution in thought of a new thing for the old. We agree that a thing is always replaced by another thing, and even that our mind cannot think the disappearance of an object, external or internal, without thinking—under an indeterminate and confused form, it is true—that another object is substituted for it. But we add that the representation of a disappearance is that of a phenomenon that is produced in space or at least in time, that consequently it still implies the calling up of an image, and that it is precisely here that we have to free ourselves from the imagination in order to appeal to the pure understanding. "Let us therefore no longer speak," it will be said, "of disappearance or annihilation; these are physical operations. Let us no longer represent the object A as annihilated or absent. Let us say simply that we think it "non-existent." To annihilate it is to act on it in time and perhaps also in space; it is to accept, consequently, the condition of spatial and temporal existence, to accept the universal connection that binds an object to all others, and prevents it from disappearing without being at the same time replaced. But we can free ourselves from these conditions; all that is necessary is that by an effort of abstraction we should call up the idea of the object A by itself, that we should agree first to consider it as existing, and then, by a stroke of the intellectual pen, blot out the clause. The object will then be, by our decree, "non-existent."

Very well, let us strike out the clause. We must not suppose that our pen-stroke is self-sufficient—that it can be isolated from the rest of things. We shall see that it carries with it, whether we will or no, all that we tried to abstract from. Let us compare together the two ideas—the object A supposed to exist, and the same object supposed "non-existent."

The idea of the object A, supposed existent, is the representation pure and simple of the object A, for we cannot represent an object without attributing to it, by the very fact of representing it, a certain reality. Between thinking an object and thinking it existent, there is absolutely no difference. Kant has put this point in clear light in his criticism of the ontological argument. Then, what is it to think the object A non-existent? To represent it non-existent cannot consist in withdrawing from the idea of the object A the idea of the attribute "existence," since, I repeat, the representation of the existence of the object is inseparable from the representation of the object, and indeed is one with it. To represent the object A non-existent can only consist, therefore, in *adding* something to the idea of this object: we add to it, in fact, the idea of an *exclusion* of this particular object by actual reality in general. To think the object A as non-existent is first to think the object and consequently to think it existent; it is then to think that another reality, with which it is incompatible, supplants it. Only, it is useless to represent this latter reality explicitly; we are not concerned with what it is; it is enough for us to know that it drives out the object A, which alone is of interest to us. That is why we think of the expulsion rather than of the cause which expels. But this cause is none the less present to the mind; it is there in the implicit state, that which expels being inseparable from the expulsion as the hand which drives the pen is inseparable from the pen-stroke. The act by which we declare an object unreal therefore posits the existence of the real in general. In other words, to represent an object as unreal cannot consist in depriving it of every kind of existence, since the representation of an object is necessarily that of the object existing. Such an act consists simply in declaring that the existence attached by our mind to the object, and inseparable from its representation, is an existence wholly ideal—that of a mere *possible*. But the "ideality" of an object, and the "simple possibility" of an object, have meaning only in relation to a reality that drives into the region of the ideal, or of the merely possible, the object which is incompatible with it. Suppose the stronger and more substantial existence annihilated: it is the attenuated and weaker existence of the merely possible that becomes the reality itself, and you will no longer be representing the object, then, as non-existent. In other words, and however strange our assertion may seem, *there is* more, *and not* less, *in the idea of an object conceived as "not existing" than in the idea of this same object conceived as "existing"; for the idea of the object "not existing" is necessarily the idea of the object "existing" with, in addition, the representation of an exclusion of this object by the actual reality taken in block.*

But it will be claimed that our idea of the non-existent is not yet sufficiently cut loose from every imaginative element, that it is not negative enough. "No matter," we shall be told, "though the unreality of a thing consist in its exclusion by other things; we want to know nothing about that. Are we not free to direct our attention where we please and how we please? Well then, after having called up the idea of an object, and thereby, if you will have it so, supposed it existent, we shall merely couple to our affirmation a 'not,' and that will be enough to make us think it non-existent. This is an operation entirely intellectual, independent of what happens outside the mind. So let us think of anything or let us think of the totality of things, and then write in the margin of our thought the 'not,' which prescribes the rejection of what it contains: we annihilate everything mentally by the mere fact of decreeing its annihilation."—Here we have it! The very root of all the difficulties and errors with which we are confronted is to be found in the power ascribed here to negation. We represent negation as exactly symmetrical with affirmation. We imagine that negation, like affirmation, is self-sufficient. So that negation, like affirmation, would have the power of creating ideas, with this sole difference that they would be negative ideas. By affirming one thing, and then another, and so on *ad infinitum,* I form the idea of "All"; so, by denying one thing and then other things, finally by denying All, I arrive at the idea of Nothing.—But it is just this assimilation which is arbitrary. We fail to see that while affirmation is a complete act of the mind, which can succeed in building up an idea, negation is but the half of an intellectual act, of which the other half is understood, or rather put off to an indefinite future. We fail to see that while affirmation is a purely intellectual act, there enters into negation an element which is not intellectual, and that it is precisely to the intrusion of this foreign element that negation owes its specific character.

To begin with the second point, let us note that to deny always consists in setting aside a possible affirmation.[1] Negation is only an attitude taken by the mind toward an eventual affirmation. When I say, "This table is

[1] Kant, *Critique of Pure Reason*, 2nd edition, p. 737: "From the point of view of our knowledge in general . . . the peculiar function of negative propositions is simply to prevent error." Cf. Sigwart, *Logik*, 2nd edition, vol. i. pp. 150 ff.

black," I am speaking of the table; I have seen it black, and my judgment expresses what I have seen. But if I say, "This table is not white," I surely do not express something I have perceived, for I have seen black, and not an absence of white. It is therefore, at bottom, not on the table itself that I bring this judgment to bear, but rather on the judgment that would declare the table white. I judge a judgment and not the table. The proposition, "This table is not white," implies that you might believe it white, that you did believe it such, or that I was going to believe it such. I warn you or myself that this judgment is to be replaced by another (which, it is true, I leave undetermined). Thus, while affirmation bears directly on the thing, negation aims at the thing only indirectly, through an interposed affirmation. An affirmative proposition expresses a judgment on an object; a negative proposition expresses a judgment on a judgment. *Negation, therefore, differs from affirmation properly so called in that it is an affirmation of the second degree: it affirms something of an affirmation which itself affirms something of an object.*

But it follows at once from this that negation is not the work of pure mind, I should say of a mind placed before objects and concerned with them alone. When we deny, we give a lesson to others, or it may be to ourselves. We take to task an interlocutor, real or possible, whom we find mistaken and whom we put on his guard. He was affirming something: we tell him he ought to affirm something else (though without specifying the affirmation which must be substituted). There is no longer then, simply, a person and an object; there is, in face of the object, a person speaking to a person, opposing him and aiding him at the same time; there is a beginning of society. Negation aims at someone, and not only, like a purely intellectual operation, at something. It is of a pedagogical and social nature. It sets straight or rather warns, the person warned and set straight being possibly, by a kind of doubling, the very person that speaks.

So much for the second point; now for the first. We said that negation is but the half of an intellectual act, of which the other half is left indeterminate. If I pronounce the negative proposition, "This table is not white," I mean that you ought to substitute for your judgment, "The table is white," another judgment. I give you an admonition, and the admonition refers to the necessity of a substitution. As to what you ought to substitute for your affirmation, I tell you nothing, it is true. This may be because I do not know the color of the table; but it is also, it is indeed even more, because the white color is that alone that interests us for the moment, so that I only need to tell you that some other color will have to be substituted for white, without having to say which. A negative judgment is therefore really one which indicates a need of substituting for an affirmative judgment another affirmative judgment, the nature of which, however, is not specified, sometimes because it is not known, more often because it fails to offer any actual interest, the attention bearing only on the substance of the first.

Thus, whenever I add a "not" to an affirmation, whenever I deny, I perform two very definite acts: (1) I interest myself in what one of my fellow-men affirms, or in what he was going to say, or in what might have been said by another *Me*, whom I anticipate; (2) I announce that some other affirmation, whose content I do not specify, will have to be substituted for the one I find before me. Now, in neither of these two acts is there anything but affirmation. The *sui generis* character of negation is due to superimposing the first of these acts upon the second. It is in vain, then, that we attribute to negation the power of creating ideas *sui generis*, symmetrical with those that affirmation creates, and directed in a contrary sense. No idea will come forth from negation, for it has no other content than that of the affirmative judgment which it judges.

To be more precise, let us consider an existential, instead of an attributive, judgment. If I say, "The object A does not exist," I mean by that, first, that we might believe that the object A exists: how, indeed, can we think of the object A without thinking it existing, and, once again, what difference can there be between the idea of the object A existing and the idea pure and simple of the object A? Therefore, merely by saying "The object A," I attribute to it some kind of existence, though it be that of a mere *possible*, that is to say, of a pure idea. And consequently, in the judgment "The object A is not," there is at first an affirmation such as "The object A has been," or "The object A will be," or, more generally, "The object A exists at least as a mere *possible*." Now, when I add the two words "is not," I can only mean that if we go further, if we erect the possible object into a real object, we shall be mistaken, and that the possible of which I am speaking is excluded from the actual reality as incompatible with it. Judgments that posit the nonexistence of a thing are therefore judgments that formulate a contrast between the possible and the actual (that is, between two kinds of *existence*, one thought and the other found), where a person, real or imaginary, wrongly believes that a certain possible is realized. Instead of this possible, there is a reality that differs from it and rejects it: the negative judgment expresses this contrast, but it expresses the contrast in an intentionally incomplete form, because it is addressed to a person who is supposed to be interested exclusively in the possible that is indicated, and is not concerned to know by what kind of reality the possible is replaced. The expression of the substitution is therefore bound to be cut short. Instead of affirming that a second term is substituted for the first, the attention which was originally directed to the first term will be kept fixed upon it, and upon it alone. And, without going beyond the first, we shall implicitly affirm that a second term replaces it in saying that the first "is not." We shall thus judge a judgment instead of judging a thing. We shall warn others or warn ourselves of a possible error instead of supplying positive information. Suppress every intention of this kind, give knowledge back its exclusively scientific or philosophical character, suppose in other words that reality comes itself to inscribe itself on a mind that cares only for things and is not interested in persons: we shall affirm that such or such a thing is, we shall never affirm that a thing is not.

How comes it, then, that affirmation and negation are so persistently put on the same level and endowed with an equal objectivity? How comes it that we have so much difficulty in recognizing that negation is subjective, artificially cut short, relative to the human mind and still more to the social life? The reason is, no doubt, that *both* negation and affirmation are expressed in propositions, and that *any* proposition, being formed of *words*, which symbolize *concepts*, is something relative to social life and to the human intellect. Whether I say "The ground is damp" or "The ground is not damp," in both cases the terms "ground" and "damp" are concepts more or less artificially created by the mind of man—extracted, by his free initiative, from the continuity of experience. In both cases the concepts are represented by the same conventional words. In both cases we can say indeed that the proposition aims at a social and pedagogical end, since the first would propagate a truth as the second would prevent an error. From this point of view, which is that of formal logic, to affirm and to deny are indeed two mutually symmetrical acts, of which the first establishes a relation of agreement and the second a relation of disagreement between a subject and an attribute. But how do we fail to see that the symmetry is altogether external and the likeness superficial? Suppose language fallen into disuse, society dissolved, every intellectual initiative, every faculty of self-reflection and of self-judgment atrophied in man: the dampness of the ground will subsist none the less, capable of inscribing itself automatically in sensation and of sending a vague idea to the deadened intellect. The intellect will still affirm, in implicit terms. And consequently, neither distinct concepts, nor words, nor the desire of spreading the truth, nor that of bettering oneself, are of the very essence of the affirmation. But this passive intelligence, mechanically keeping step with experience, neither anticipating nor following the course of the real, would have no wish to deny. It could not receive an imprint of negation; for, once again, that which exists may come to be recorded, but the non-existence of the non-existing cannot. For such an intellect to reach the point of denying, it must awake from its torpor, formulate the disappointment of a real or possible expectation, correct an actual or possible error—in short, propose to teach others or to teach itself.

It is rather difficult to perceive this in the example we have chosen, but the example is indeed the more instructive and the argument the more cogent on that account. If dampness is able automatically to come and record itself, it is the same, it will be said, with non-dampness; for the dry as well as the damp can give impressions to sense, which will transmit them, as more or less distinct ideas, to the intelligence. In this sense the negation of dampness is as objective a thing, as purely intellectual, as remote from every pedagogical intention, as affirmation.—But let us look at it more closely: we shall see that the negative proposition, "The ground is not damp," and the affirmative proposition, "The ground is dry," have entirely different contents. The second implies that we know the dry, that we have experienced the specific sensations, tactile or visual for example, that are at the base of this idea. The first requires nothing of the sort; it could equally well have been formulated by an intelligent fish, who had never perceived anything but the wet. It would be necessary, it is true, that this fish should have risen to the distinction between the real and the possible, and that he should care to anticipate the error of his fellow-fishes, who doubtless consider as alone possible the condition of wetness in which they actually live. Keep strictly to the terms of the proposition, "The ground is not damp," and you will find that it means two things: (1) that one might believe that the ground is damp, (2) that the dampness is replaced in fact by a certain quality x. This quality is left indeterminate, either because we have no positive knowledge of it, or because it has no actual interest for the person to whom the negation is addressed. To deny, therefore, always consists in presenting in an abridged form a system of two affirmations: the one determinate, which applies to a certain *possible*; the other indeterminate, referring to the unknown or indifferent reality that supplants this possibility. The second affirmation is virtually contained in the judgment we apply to the first, a judgment which is negation itself. And what gives negation its subjective character is precisely this, that in the discovery of a replacement it takes account only of the replaced, and is not concerned with what replaces. The replaced exists only as a conception of the mind. It is necessary, in order to continue to see it, and consequently in order to speak of it, to turn our back on the reality, which flows from the past to the present, advancing from behind. It is this that we do when we deny. We discover the change, or more generally the substitution, as a traveler would see the course of his carriage if he looked out behind, and only knew at each moment the point at which he had ceased to be; he could never determine his actual position except by relation to that which he had just quitted, instead of grasping it in itself.

To sum up, for a mind which should follow purely and simply the thread of experience, there would be no void, no nought, even relative or partial, no possible negation. Such a mind would see facts succeed facts, states succeed states, things succeed things. What it would note at each moment would be things existing, states appearing, events happening. It would live in the actual, and, if it were capable of judging, it would never affirm anything except the existence of the present.

Endow this mind with memory, and especially with the desire to dwell on the past; give it the faculty of dissociating and of distinguishing: it will no longer only note the present state of the passing reality; it will represent the passing as a change, and therefore as a contrast between what has been and what is. And as there is no essential difference between a past that we remember and a past that we imagine, it will quickly rise to the idea of the "possible" in general.

It will thus be shunted on to the siding of negation. And especially it will be at the point of representing a disappearance. But it will not yet have reached it. To represent that a thing has disappeared, it is not enough to perceive a contrast between the past and the present; it is necessary besides to turn our back on the present, to dwell on the past, and to think the contrast of the past with the present in terms of the past only, without letting the present appear in it.

The idea of annihilation is therefore not a pure idea; it implies that we regret the past or that we conceive it as regrettable, that we have some reason to linger over it. The idea arises when the phenomenon of substitution is cut in two by a mind which considers only the first half, because that alone interests it. Suppress all interest, all feeling, and there is nothing left but the reality that flows, together with the knowledge ever renewed that it impresses on us of its present state.

From annihilation to negation, which is a more general operation, there is now only a step. All that is necessary is to represent the contrast of what is, not only with what has been, but also with all that might have been. And we must express this contrast as a function of what might have been, and not of what is; we must affirm the existence of the actual while looking only at the possible. The formula we thus obtain no longer expresses merely a disappointment of the individual; it is made to correct or guard against an error, which is rather supposed to be the error of another. In this sense, negation has a pedagogical and social character.

Now, once negation is formulated, it presents an aspect symmetrical with that of affirmation; if affirmation affirms an objective reality, it seems that negation must affirm a non-reality equally objective, and, so to say, equally real. In which we are both right and wrong: wrong, because negation cannot be objectified, in so far as it is negative; right, however, in that the negation of a thing implies the latent affirmation of its replacement by something else, which we systematically leave on one side. But the negative form of negation benefits by the affirmation at the bottom of it. Bestriding the positive solid reality to which it is attached, this phantom objectifies itself. Thus is formed the idea of the void or of a partial nought, a thing being supposed to be replaced, not by another thing, but by a void which it leaves, that is, by the negation of itself. Now, as this operation works on anything whatever, we suppose it performed on each thing in turn, and finally on all things in block. We thus obtain the idea of absolute Nothing. If now we analyze this idea of Nothing, we find that it is, at bottom, the idea of Everything, together with a movement of the mind that keeps jumping from one thing to another, refuses to stand still, and concentrates all its attention on this refusal by never determining its actual position except by relation to that which it has just left. It is therefore an idea eminently comprehensive and full, as full and comprehensive as the idea of *All*, to which it is very closely akin.

How then can the idea of Nought be opposed to that of All? Is it not plain that this is to oppose the full to the full, and that the question, "Why does something exist?" is consequently without meaning, a pseudo-problem raised about a pseudo-idea? Yet we must say once more why this phantom of a problem haunts the mind with such obstinacy. In vain do we show that in the idea of an "annihilation of the real" there is only the image of all realities expelling one another endlessly, in a circle; in vain do we add that the idea of non-existence is only that of the expulsion of an imponderable existence, or a "merely possible" existence, by a more substantial existence which would then be the true reality; in vain do we find in the *sui generis* form of negation an element which is not intellectual—negation being the judgment of a judgment, an admonition given to someone else or to oneself, so that it is absurd to attribute to negation the power of creating ideas of a new kind, viz. ideas without content;—in spite of all, the conviction persists that before things, or at least under things, there is "Nothing." If we seek the reason of this fact, we shall find it precisely in the feeling, in the social and, so to speak, practical element, that gives its specific form to negation. The greatest philosophic difficulties arise, as we have said, from the fact that the forms of human action venture outside of their proper sphere. We are made in order to act as much as, and more than, in order to think—or rather, when we follow the bent of our nature, it is in order to act that we think. It is therefore no wonder that the habits of action give their tone to those of thought, and that our mind always perceives things in the same order in which we are accustomed to picture them when we propose to act on them. Now, it is unquestionable, as we remarked above, that every human action has its starting-point in a dissatisfaction, and thereby in a feeling of absence. We should not act if we did not set before ourselves an end, and we seek a thing only because we feel the lack of it. Our action proceeds thus from "nothing" to "something," and its very essence is to embroider "something" on the canvas of "nothing." The truth is that the "nothing" concerned here is the absence not so much of a thing as of a utility. If I bring a visitor into a room that I have not yet furnished, I say to him that "there is nothing in it." Yet I know the room is full of air; but, as we do not sit on air, the room truly contains nothing that at this moment, for the visitor and for myself, counts for anything. In a general way, human work consists in creating utility; and, as long as the work is not done, there is "nothing"—nothing that we want. Our life is thus spent in filling voids, which our intellect conceives under the influence, by no means intellectual, of desire and of regret, under the pressure of vital necessities; and if we mean by void an absence of utility and not of things, we may say, in this quite relative sense, that we are constantly going from the void to the full: such is the direction which our action takes. Our speculation cannot help doing the same; and, naturally, it passes from the relative sense to the absolute sense, since it is exercised on things themselves and not on the utility they have for us. Thus is implanted in us the idea that reality fills a void, and that Nothing, conceived as an absence of everything, pre-exists before all things in right, if not in fact. It is this illusion that we have tried to remove by showing that the idea of Nothing, if we try to see in it that of an annihilation of all things, is self-destructive and reduced to a mere word; and that if, on the contrary, it is truly an idea, then we find in it as much matter as in the idea of All.

This long analysis has been necessary to show that *a self-sufficient reality is not necessarily a reality foreign to duration.* If we pass (consciously or unconsciously) through the idea of the nought in order to reach that of being, the being to which we come is a logical or mathematical essence, therefore non-temporal. And, consequently, a static conception of the real is forced on us: everything appears given once for all, in eternity. But we must accustom ourselves to think being directly, without making a detour, without first appealing to the phantom of the nought which interposes itself between it and us. We must strive to see in order to see, and no longer to see in order to act. Then the Absolute is revealed very near us and, in a certain measure, in us. It is of psychological and not of mathematical nor logical essence. It lives with us. Like us, but in certain aspects infinitely more concentrated and more gathered up in itself, it *endures.*

But do we ever think true duration? Here again a direct taking possession is necessary. It is no use trying to approach duration: we must install ourselves within it straight away. This is what the intellect generally refuses to do, accustomed as it is to think the moving by means of the unmovable.

The function of the intellect is to preside over actions. Now, in action, it is the result that interests us; the means matter little provided the end is attained. Thence it comes that we are altogether bent on the end to be realized, generally trusting ourselves to it in order that the idea may become an act; and thence it comes also that only the goal where our activity will rest is pictured explicitly to our mind: the movements constituting the action itself either elude our consciousness or reach it only confusedly. Let us consider a very simple act, like that of lifting the arm. Where should we be if we had to imagine beforehand all the elementary contractions and tensions this act involves, or even to perceive them, one by one, as they are accomplished? But the mind is carried immediately to the end, that is to say, to the schematic and simplified vision of the act supposed accomplished. Then, if no antagonistic idea neutralizes the effect of the first idea, the appropriate movements come of themselves to fill out the plan, drawn in some way by the void of its gaps. The intellect, then, only represents to the activity ends to attain, that is to say, points of rest. And, from one end attained to another end attained, from one rest to another rest, our activity is carried by a series of leaps, during which our consciousness is turned away as much as possible from the movement going on, to regard only the anticipated image of the movement accomplished.

Now, in order that it may represent as unmovable the

result of the act which is being accomplished, the intellect must perceive, as also unmovable, the surroundings in which this result is being framed. Our activity is fitted into the material world. If matter appeared to us as a perpetual flowing, we should assign no termination to any of our actions. We should feel each of them dissolve as fast as it was accomplished, and we should not anticipate an ever-fleeting future. In order that our activity may leap from an *act* to an *act,* it is necessary that matter should pass from a *state* to a *state*, for it is only into a state of the material world that action can fit a result, so as to be accomplished. But is it thus that matter presents itself?

A priori we may presume that our perception manages to apprehend matter with this bias. Sensory organs and motor organs are in fact co-ordinated with each other. Now, the first symbolize our faculty of perceiving, as the second our faculty of acting. The organism thus evidences, in a visible and tangible form, the perfect accord of perception and action. So if our activity always aims at a *result* into which it is momentarily fitted, our perception must retain of the material world, at every moment, only a *state* in which it is provisionally placed. This is the most natural hypothesis. And it is easy to see that experience confirms it.

From our first glance at the world, before we even make our *bodies* in it, we distinguish *qualities*. Color succeeds to color, sound to sound, resistance to resistance, etc. Each of these qualities, taken separately, is a state that seems to persist as such, immovable until another replaces it. Yet each of these qualities resolves itself, on analysis, into an enormous number of elementary movements. Whether we see in it vibrations or whether we represent it in any other way, one fact is certain, it is that every quality is change. In vain, moreover, shall we seek beneath the change the thing which changes: it is always provisionally, and in order to satisfy our imagination, that we attach the movement to a mobile. The mobile flies forever before the pursuit of science, which is concerned with mobility alone. In the smallest discernible fraction of a second, in the almost instantaneous perception of a sensible quality, there may be trillions of oscillations which repeat themselves. The permanence of a sensible quality consists in this repetition of movements, as the persistence of life consists in a series of palpitations. The primal function ot perception is precisely to grasp a series of elementary changes under the form of a quality or of a simple state, by a work of condensation. The greater the power of acting bestowed upon an animal species, the more numerous, probably, are the elementary changes that its faculty of perceiving concentrates into one of its instants. And the progress must be continuous, in nature, from the beings that vibrate almost in unison with the oscillations of the ether, up to those that embrace trillions of these oscillations in the shortest of their simple perceptions. The first feel hardly anything but movements; the others perceive quality. The first are almost caught up in the running-gear of things; the others react, and the tension of their faculty of acting is probably proportional to the concentration of their faculty of perceiving. The progress goes on even in humanity itself. A man is so much the more a "man of action" as he can embrace in a glance a greater number of events: he who perceives successive events one by one will allow himself to be led by them; he who grasps them as a whole will dominate them. In short, the qualities of matter are so many stable views that we take of its instability.

Now, in the continuity of sensible qualities we mark off the boundaries of bodies. Each of these bodies really changes at every moment. In the first place, it resolves itself into a group of qualities, and every quality, as we said, consists of a succession of elementary movements. But, even if we regard the quality as a stable state, the body is still unstable in that it changes qualities without ceasing. The body pre-eminently—that which we are most justified in isolating within the continuity of matter, because it constitutes a relatively closed system—is the living body; it is, moreover, for it that we cut out the others within the whole. Now, life is an evolution. We concentrate a period of this evolution in a stable view which we call a form, and, when the change has become considerable enough to overcome the fortunate inertia of our perception, we say that the body has changed its form. But in reality the body is changing form at every moment; or rather, there is no form, since form is immobile and the reality is movement. What is real is the continual *change of* form: *form is only a snapshot view of a transition.* Therefore, here again, our perception manages to solidify into discontinuous images the fluid continuity of the real. When the successive images do not differ from each other too much, we consider them all as the waxing and waning of a single *mean* image, or as the deformation of this image in different directions. And to this mean we really allude when we speak of the *essence* of a thing, or of the thing itself.

Finally things, once constituted, show on the surface, by their changes of situation, the profound changes that are being accomplished within the Whole. We say then that they *act* on one another. This action appears to us, no doubt, in the form of movement. But from the mobility of the movement we turn away as much as we can; what interests us is, as we said above, the unmovable plan of the movement rather than the movement itself. Is it a simple movement? We ask ourselves *where* it is going. It is by its direction, that is to say, by the position of its provisional end, that we represent it at every moment. Is it a complex movement? We would know above all *what* is going on, *what* the movement is doing—in other words, the *result* obtained or the presiding *intention*. Examine closely what is in your mind when you speak of an action in course of accomplishment. The idea of change is there, I am willing to grant, but it is hidden in the penumbra. In the full light is the motionless plan of the act supposed accomplished. It is by this, and by this only, that the complex act is distinguished and defined. We should be very much embarrassed if we had to imagine the movements inherent in the actions of eating, drinking, fighting, etc. It is enough for us to know, in a general and indefinite way, that all these acts are movements. Once that side of the matter has been settled, we simply seek to represent the *general plan* of each of these complex movements, that is to say the *motionless design* that underlies them. Here again knowledge bears on a state rather than on a change. It is therefore the same with this third case as with the others. Whether the movement be qualitative or evolutionary or extensive, the mind manages to take stable views of the instability.

And thence the mind derives, as we have just shown, three kinds of representations: (1) qualities, (2) forms of essences, (3) acts.

To these three ways of seeing correspond three categories of words: *adjectives, substantives* and *verbs,* which are the primordial elements of language. Adjectives and substantives therefore symbolize *states.* But the verb itself, if we keep to the clear part of the idea it calls up, hardly expresses anything else.

Now, if we try to characterize more precisely our natural attitude toward Becoming, this is what we find. Becoming is infinitely varied. That which goes from yellow to green is not like that which goes from green to blue: they are different *qualitative* movements. That which goes from flower to fruit is not like that which goes from larva to nymph and from nymph to perfect insect: they are different *evolutionary* movements. The action of eating or of drinking is not like the action of fighting: they are different *extensive* movements. And these three kinds of movement themselves—qualitative, evolutionary, extensive—differ profoundly. The trick of our perception, like that of our intelligence, like that of our language, consists in extracting from these profoundly different becomings the single representation of becoming *in general,* undefined becoming, a mere abstraction which by itself says nothing and of which, indeed, it is very rarely that we think. To this idea, always the same, and always obscure or unconscious, we then join, in each particular case, one or several clear images that represent *states* and which serve to distinguish all becomings from each other. It is this composition of a specified and definite state with change general and undefined that we substitute for the specific change. An infinite multiplicity of becomings variously colored, so to speak, passes before our eyes: we manage so that we see only differences of color, that is to say, differences of state, beneath which there is supposed to flow, hidden from our view, a becoming always and everywhere the same, invariably colorless.

Suppose we wish to portray on a screen a living picture, such as the marching past of a regiment. There is one way in which it might first occur to us to do it. That would be to cut out jointed figures representing the soldiers, to give to each of them the movement of marching, a movement varying from individual to individual although common to the human species, and to throw the whole on the screen. We should need to spend on this little game an enormous amount of work, and even then we should obtain but a very poor result: how could it, at its best, reproduce the suppleness and variety of life? Now, there is another way of proceeding, more easy and at the same time more effective. It is to take a series of snapshots of the passing regiment and to throw these instantaneous views on the screen, so that they replace each other very rapidly. This is what the cinematograph does. With photographs, each of which represents the regiment in a fixed attitude, it reconstitutes the mobility of the regiment marching. It is true that if we had to do with photographs alone, however much we might look at them, we should never see them animated: with immobility set beside immobility, even endlessly, we could never make movement. In order that the pictures may be animated, there must be movement somewhere. The movement does indeed exist here; it is in the apparatus. It is because the film of the cinematograph unrolls, bringing in turn the different photographs of the scene to continue each other, that each actor of the scene recovers his mobility; he strings all his successive attitudes on the invisible movement of the film. The process then consists in extracting from all the movements peculiar to all the figures an impersonal movement abstract and simple, *movement in general*, so to speak: we put this into the apparatus, and we reconstitute the individuality of each particular movement by combining this nameless movement with the personal attitudes. Such is the contrivance of the cinematograph. And such is also that of our knowledge. Instead of attaching ourselves to the inner becoming of things, we place ourselves outside them in order to recompose their becoming artificially. We take snapshots, as it were, of the passing reality, and, as these are characteristic of the reality, we have only to string them on a becoming, abstract, uniform and invisible, situated at the back of the apparatus of knowledge, in order to imitate what there is that is characteristic in this becoming itself. Perception, intellection, language so proceed in general. Whether we would think becoming, or express it, or even perceive it, we hardly do anything else than set going a kind of cinematograph inside us. We may therefore sum up what we have been saying in the conclusion that the *mechanism of our ordinary knowledge is of a cinematographical kind.*

Of the altogether practical character of this operation there is no possible doubt. Each of our acts aims at a certain insertion of our will into the reality. There is, between our body and other bodies, an arrangement like that of the pieces of glass that compose a kaleidoscopic picture. Our activity goes from an arrangement to a re-arrangement, each time no doubt giving the kaleidoscope a new shake, but not interesting itself in the shake, and seeing only the new picture. Our knowledge of the operation of nature must be exactly symmetrical, therefore, with the interest we take in our own operation. In this sense we may say, if we are not abusing this kind of illustration, that *the cinematographical character of our knowledge of things is due to the kaleidoscopic character of our adaptation to them.*

The cinematographical method is therefore the only practical method, since it consists in making the general character of knowledge form itself on that of action, while expecting that the detail of each act should depend in its turn on that of knowledge. In order that action may always be enlightened, intelligence must always be present in it; but intelligence, in order thus to accompany the progress of activity and ensure its direction, must begin by adopting its rhythm. Action is discontinuous, like every pulsation of life; discontinuous, therefore, is knowledge. The mechanism of the faculty of knowing has been constructed on this plan. Essentially practical, can it be of use, such as it is, for speculation? Let us try with it to follow reality in its windings, and see what will happen.

I take of the continuity of a particular becoming a series of views, which I connect together by "becoming in general." But of course I cannot stop there. What is not determinable is not representable: of "becoming in general" I have only a verbal knowledge. As the letter x designates a certain unknown quantity, whatever it may be, so my "becoming in general," always the same, symbolizes here a certain transition of which I have taken some snapshots; of the transition itself it teaches me nothing. Let me then concentrate myself wholly on the transition, and, between any two snapshots, endeavor to realize what is going on. As I apply the same method, I obtain the same result; a third view merely slips in between the two others. I may begin again as often as I will, I may set views alongside of views for ever, I shall obtain nothing else. The application of the cinematographical method therefore leads to a perpetual recommencement, during which the mind, never able to satisfy itself and never finding where to rest, persuades itself, no doubt, that it imitates by its instability the very movement of the real. But though, by straining itself to the point of giddiness, it may end by giving itself the illusion of mobility, its operation has not advanced it a step, since it remains as far as ever from its goal. In order to advance with the moving reality, you must replace yourself within it. Install yourself within change, and you will grasp at once both change itself and the successive states in which *it might* at any instant be immobilized. But with these successive states, perceived from without as real and no longer as potential immobilities, you will never reconstitute movement. Call them *qualities*, *forms*, *positions*, or *intentions*, as the case may be, multiply the number of them as you will, let the interval between two consecutive states be infinitely small: before the intervening movement you will always experience the disappointment of the child who tries by clapping his hands together to crush the smoke. The movement slips through the interval, because every attempt to reconstitute change out of states implies the absurd proposition, that movement is made of immobilities.

Philosophy perceived this as soon as it opened its eyes. The arguments of Zeno of Elea, although formulated with a very different intention, have no other meaning.

Take the flying arrow. At every moment, says Zeno, it is motionless, for it cannot have time to move, that is, to occupy at least two successive positions, unless at least two moments are allowed it. At a given moment, therefore, it is at rest at a given point. Motionless in each point of its course, it is motionless during all the time that it is moving.

Yes, if we suppose that the arrow can ever *be* in a point of its course. Yes again, if the arrow, which is moving, ever coincides with a position, which is motionless. But the arrow never *is* in any point of its course. The most we can say is that it might be there, in this sense, that it passes there and might stop there. It is true that if it did stop there, it would be at rest there, and at this point it is no longer movement that we should have to do with. The truth is that if the arrow leaves the point A to fall down at the point B, its movement AB is as simple, as indecomposable, in so far as it is movement, as the tension of the bow that shoots it. As the shrapnel, bursting before it falls to the ground, covers the explosive zone with an indivisible danger, so the arrow which goes from A to B displays with a single stroke, although over a certain extent of duration, its indivisible mobility. Suppose an elastic stretched from A to B, could you divide its extension? The course of the arrow is this very extension; it is equally simple and equally undivided. It is a single and unique bound. You fix a point C in the interval passed, and say that at a certain moment the arrow was in C. If it had been there, it would have been stopped there, and you would no longer have had a flight from A to B, but *two* flights, one from A to C and the other from C to B, with an interval of rest. A single movement is entirely, by the hypothesis, a movement between two stops; if there are intermediate stops, it is no longer a single movement. At bottom, the illusion arises from this, that the movement, *once effected*, has laid along its course a motionless trajectory on which we can count as many immobilities as we will. From this we conclude that the movement, *whilst being effected*, lays at each instant beneath it a position with which it coincides. We do not see that the trajectory is created in one stroke, although a certain time is required for it; and that though we can divide at will the trajectory once created, we cannot divide its creation, which is an act in progress and not a thing. To suppose that the moving body *is* at a point of its course is to cut the course in two by a snip of the scissors at this point, and to substitute two trajectories for the single trajectory which we were first considering. It is to distinguish two successive acts where, by the hypothesis, there is only one. In short, it is to attribute to the course itself of the arrow everything that can be said of the interval that the arrow has traversed, that is to say, to admit *a priori* the absurdity that movement coincides with immobility.

We shall not dwell here on the three other arguments of Zeno. We have examined them elsewhere. It is enough to point out that they all consist in applying the movement to the line traversed, and supposing that what is true of the line is true of the movement. The line, for example, may be divided into as many parts as we wish, of any length that we wish, and it is always the same line. From this we conclude that we have the right tc suppose the movement articulated as we wish, and that it is always the same movement. We thus obtain a series of absurdities that all express the same fundamental absurdity. But the possibility of applying the movement *to* the line traversed exists only for an observer who, keeping outside the movement and seeing at every instant the possibility of a stop, tries to reconstruct the real movement with these possible immobilities. The absurdity vanishes as soon as we adopt by thought the continuity of the real movement, a continuity of which every one of us is conscious whenever he lifts an arm or advances a step. We feel then indeed that the line passed over between two stops is described with a single indivisible stroke, and that we seek in vain to practice on the movement, which traces the line, divisions corresponding, each to each, with the divisions arbitrarily chosen of the line once it has been traced. The line traversed by the moving body lends itself to any kind of division, because it has no internal organization. But all movement is articulated inwardly. It is either an indivisible bound (which may occupy, nevertheless, a very long duration) or a series of indivisible bounds. Take the articulations of this movement into account, or give up speculating on its nature.

When Achilles pursues the tortoise, each of his steps must be treated as indivisible, and so must each step of the tortoise. After a certain number of steps, Achilles will have overtaken the tortoise. There is nothing more simple. If you insist on dividing the two motions further, distinguish both on the one side and on the other, in the course of Achilles and in that of the tortoise, the *sub-multiples* of the steps of each of them; but respect the natural articulations of the two courses. As long as you respect them, no difficulty will arise, because you will follow the indications of experience. But Zeno's device is to reconstruct the movement of Achilles according to a law arbitrarily chosen. Achilles with a first step is supposed to arrive at the point where the tortoise was, with a second step at the point which it has moved to while he was making the first, and so on. In this case, Achilles would always have a new step to take. But obviously, to overtake the tortoise, he goes about it in quite another way. The movement considered by Zeno would only be the equivalent of the movement of Achilles if we could treat the movement as we treat the interval passed through, decomposable and recomposable at will. Once you subscribe to this first absurdity, all the others follow.[1]

[1] That is, we do not consider the sophism of Zeno refuted by the fact that the geometrical progression $a\ (1 + \frac{1}{n} + \frac{1}{n^2} + \frac{1}{n^3} + \ldots$, etc.)—in which a designates the initial distance between Achilles and the tortoise, and n the relation of their respective velocities—has a finite sum if n is greater than 1. On this point we may refer to the arguments of F. Evellin, which we regard as conclusive (see Evellin, *Infini et quantité*, Paris, 1880, pp. 63-97; cf. *Revue philosophique*, vol. xi., 1881, pp. 564-568). The truth is that mathematics, as we have tried to show in a former work, deals and can deal only with lengths. It has therefore had to seek devices, first, to transfer to the movement, which is not a length, the divisibility of the line passed over, and then to reconcile with experience the idea (contrary to experience and full of absurdities) of a movement that is a length, that is, of a movement *placed upon* its trajectory and arbitrarily decomposable like it.

Nothing would be easier, now, than to extend Zeno's argument to qualitative becoming and to evolutionary becoming. We should find the same contradictions in these. That the child can become a youth, ripen to maturity and decline to old age, we understand when we consider that vital evolution is here the reality itself. Infancy, adolescence, maturity, old age, are mere views of the mind, *possible stops* imagined by us, from without, along the continuity of a progress. On the contrary, let childhood, adolescence, maturity and old age be given as integral parts of the evolution, they become *real stops*, and we can no longer conceive how evolution is possible, for rests placed beside rests will never be equivalent to a movement. How, with what is made, can we reconstitute what is being made? How, for instance, from childhood once posited as a *thing*, shall we pass to adolescence, when, by the hypothesis, childhood only is given? If we look at it closely, we shall see that our habitual manner of speaking, which is fashioned after our habitual manner of thinking, leads us to actual logical deadlocks—deadlocks to which we allow ourselves to be led without anxiety, because we feel confusedly that we can always get out of them if we like: all that we have to do, in fact, is to give up the cinematographical habits of our intellect. When we say "The child becomes a man," let us take care not to fathom too deeply the literal meaning of the expression, or we shall find that, when we posit the subject "child," the attribute "man" does not yet apply to it, and that, when we express the attribute "man," it applies no more to the subject "child." The reality, which is the *transition* from childhood to manhood, has slipped between our fingers. We have only the imaginary stops "child" and "man," and we are very near to saying that one of these stops *is* the other, just as the arrow of Zeno *is*, according to that philosopher, at all the points of the course. The truth is that if language here were molded on reality, we should not say "The child becomes the man," but "There is becoming from the child to the man." In the first proposition, "becomes" is a verb of indeterminate meaning, intended to mask the absurdity into which we fall when we attribute the state "man" to the subject "child." It behaves in much the same way as the movement, always the same, of the cinematographical film, a movement hidden in the apparatus and whose function it is to superpose the successive pictures on one another in order to imitate the movement of the real object. In the second proposition, "becoming" is a subject. It comes to the front. It is the reality itself; childhood and manhood are then only possible stops, mere views of the mind; we now have to do with the objective movement itself, and no longer with its cinematographical imitation. But the first manner of expression is alone conformable to our habits of language. We must, in order to adopt the second, escape from the cinematographical mechanism of thought.

We must make complete abstraction of this mechanism, if we wish to get rid at one stroke of the theoretical absurdities that the question of movement raises. All is obscure, all is contradictory when we try, with states, to build up a transition. The obscurity is cleared up, the contradiction vanishes, as soon as we place ourselves along the transition, in order to distinguish states in it by making cross cuts therein in thought. The reason is that there is *more* in the transition than the series of states, that is to say, the possible cuts—*more* in the movement than the series of positions, that is to say, the possible stops. Only, the first way of looking at things is conformable to the processes of the human mind; the second requires, on the contrary, that we reverse the bent of our intellectual habits. No wonder, then, if philosophy at first recoiled before such an effort. The Greeks trusted to nature, trusted the natural propensity of the mind, trusted language above all, in so far as it naturally externalizes thought. Rather than lay blame on the attitude of thought and language toward the course of things, they preferred to pronounce the course of things itself to be wrong.

Such, indeed, was the sentence passed by the philosophers of the Eleatic school. And they passed it without any reservation whatever. As becoming shocks the habits of thought and fits ill into the molds of language, they declared it unreal. In spatial movement and in change in general they saw only pure illusion. This conclusion could be softened down without changing the premises, by saying that the reality changes, but that it *ought not* to change. Experience confronts us with becoming: that is *sensible* reality. But the *intelligible* reality, that which *ought* to be, is more real still, and that reality does not change. Beneath the qualitative becoming, beneath the evolutionary becoming, beneath the extensive becoming, the mind must seek that which defies change, the definable quality, the form or essence, the end. Such was the fundamental principle of the philosophy which developed throughout the classic age, the philosophy of Forms, or, to use a term more akin to the Greek, the philosophy of Ideas.

The word εἶδος, which we translate here by "Idea," has, in fact, this threefold meaning. It denotes (1) the quality, (2) the form or essence, (3) the end or *design* (in the sense of *intention*) of the act being performed, that is to say, at bottom, the *design* (in the sense of *drawing*) of the act supposed accomplished. *These three aspects are those of the adjective, substantive and verb, and correspond to the three essential categories of language.* After the explanations we have given above, we might, and perhaps we ought to, translate εἶδος by "view" or rather by "moment." For εἶδος is the stable view taken of the instability of things: the *quality*, which is a moment of becoming; the *form*, which is a moment of evolution; the *essence*, which is the mean form above and below which the other forms are arranged as alterations of the mean; finally, the intention or *mental design* which presides over the action being accomplished, and which is nothing else, we said, than the *material design*, traced out and contemplated beforehand, of the action accomplished. To reduce things to Ideas is therefore to resolve becoming into its principal moments, each of these being, moreover, by the hypothesis, screened from the laws of time and, as it were, plucked out of eternity. That is to say that we end in the philosophy of Ideas when we apply the cinematographical mechanism of the intellect to the analysis of the real.

But, when we put immutable Ideas at the base of the moving reality, a whole physics, a whole cosmology, a whole theology follows necessarily. We must insist on the point. Not that we mean to summarize in a few pages a philosophy so complex and so comprehensive as that of the Greeks. But, since we have described the cinematographical mechanism of the intellect, it is important that we should show to what idea of reality the play of this mechanism leads. It is the very idea, we believe, that we find in the ancient philosophy. The main lines of the doctrine that was developed from Plato to Plotinus, passing through Aristotle (and even, in a certain measure, through the Stoics), have nothing accidental, nothing contingent, nothing that must be regarded as a philosopher's fancy. They indicate the vision that a systematic intellect obtains of the universal becoming when regarding it by means of snapshots, taken at intervals, of its flowing. So that, even today, we shall philosophize in the manner of the Greeks, we shall rediscover, without needing to know them, such and such of their general conclusions, in the exact proportion that we trust in the cinematographical instinct of our thought.

We said there is *more* in a movement than in the successive positions attributed to the moving object, *more* in a becoming than in the forms passed through in turn, *more* in the evolution of form than the forms assumed one after another. Philosophy can therefore derive terms of the second kind from those of the first, but not the first from the second: from the first terms speculation must take its start. But the intellect reverses the order of the two groups; and, on this point, ancient philosophy proceeds as the intellect does. It installs itself in the immutable, it posits only Ideas. Yet becoming exists: it is a fact. How, then, having posited immutability alone, shall we make change come forth from it? Not by the addition of anything, for, by the hypothesis, there exists nothing positive outside Ideas. It must therefore be by a diminution. So at the base of ancient philosophy lies necessarily this postulate: that there is more in the motionless than in the moving, and that we pass from immutability to becoming by way of diminution or attenuation.

It is therefore something negative, or zero at most, that must be added to Ideas to obtain change. In that consists the Platonic "non-being," the Aristotelian "matter"—a metaphysical zero which, joined to the Idea, like the arithmetical zero to unity, multiplies it in space and time. By it the motionless and simple Idea is refracted into a movement spread out indefinitely. In right, there ought to be nothing but immutable Ideas, immutably fitted to each other. In fact, matter comes to add to them its void, and thereby lets loose the universal becoming. It is an elusive nothing, that creeps between the Ideas and creates endless agitation, eternal disquiet, like a suspicion insinuated between two loving hearts. Degrade the immutable Ideas: you obtain, by that alone, the perpetual flux of things. The Ideas or Forms are the whole of intelligible reality, that is to say, of truth, in that they represent, all together, the theoretical equilibrium of Being. As to sensible reality, it is a perpetual oscillation from one side to the other of this point of equilibrium.

Hence, throughout the whole philosophy of Ideas there is a certain conception of duration, as also of the relation of time to eternity. He who installs himself in becoming sees in duration the very life of things, the fundamental reality. The Forms, which the mind isolates and stores up in concepts, are then only snapshots of the changing reality. They are moments gathered along the course of time; and, just because we have cut the thread that binds them to time, they no longer endure. They tend to withdraw into their own definition, that is to say, into the artificial reconstruction and symbolical expression which is their intellectual equivalent. They enter into eternity, if you will; but what is eternal in them is just what is unreal. On the contrary, if we treat becoming by the cinematographical method, the Forms are no longer snapshots taken of the change, they are its constitutive elements, they represent all that is positive in Becoming. Eternity no longer hovers over time, as an abstraction; it underlies time, as a reality. Such is exactly, on this point, the attitude of the philosophy of Forms or Ideas. It establishes between eternity and time the same relation as between a piece of gold and the small change—change so small that payment goes on forever without the debt being paid off. The debt could be paid at once with the piece of gold. It is this that Plato expresses in his magnificent language when he says that God, unable to make the world eternal, gave it Time, "a moving image of eternity." [1]

Hence also arises a certain conception of extension, which is at the base of the philosophy of Ideas, although it has not been so explicitly brought out. Let us imagine a mind placed alongside becoming, and adopting its movement. Each successive state, each quality, each form, in short, will be seen by it as a mere cut made by thought in the universal becoming. It will be found that form is essentially extended, inseparable as it is from the extensity of the becoming which has materialized it in the course of its flow. Every form thus occupies space,

[1] Plato, *Timaeus*, 37 D.

as it occupies time. But the philosophy of Ideas follows the inverse direction. It starts from the Form; it sees in the Form the very essence of reality. It does not take Form as a snapshot of becoming; it posits Forms in the eternal; of this motionless eternity, then, duration and becoming are supposed to be only the degradation. Form thus posited, independent of time, is then no longer what is found in a perception; it is a *concept*. And, as a reality of the conceptual order occupies no more of extension than it does of duration, the Forms must be stationed outside space as well as above time. Space and time have therefore necessarily, in ancient philosophy, the same origin and the same value. The same diminution of being is expressed both by extension in space and detention in time. Both of these are but the distance between what is and what ought to be. From the standpoint of ancient philosophy, space and time can be nothing but the field that an incomplete reality, or rather a reality that has gone astray from itself, needs in order to run in quest of itself. Only it must be admitted that the field is created as the hunting progresses, and that the hunting in some way deposits the field beneath it. Move an imaginary pendulum, a mere mathematical point, from its position of equilibrium: a perpetual oscillation is started, along which points are placed next to points, and moments succeed moments. The space and time which thus arise have no more "positivity" than the movement itself. They represent the remoteness of the position artificially given to the pendulum from its normal position, *what it lacks* in order to regain its natural stability. Bring it back to its normal position: space, time and motion shrink to a mathematical point. Just so, human reasonings are drawn out into an endless chain, but are at once swallowed up in the truth seized by intuition, for their extension in space and time is only the distance, so to speak, between thought and truth.[1] So of extension and duration in relation to pure Forms or Ideas. The sensible forms are before us, ever about to recover their ideality, ever prevented by the matter they bear in them, that is to say, by their inner void, by the interval between what they are and what they ought to be. They are forever on the point of recovering themselves, forever occupied in losing themselves. An inflexible law condemns them, like the rock of Sisyphus, to fall back when they are almost touching the summit, and this law, which has projected them into space and time, is nothing other than the very constancy of their original insufficiency. The alternations of generation and decay, the evolutions ever beginning over and over again, the infinite repetition of the cycles of celestial spheres—this all represents merely a certain fundamental deficit, in which materiality consists. Fill up this deficit: at once you suppress space and time, that is to say, the endlessly renewed oscillations around a stable equilibrium always aimed at, never reached. Things re-enter into each other. What was extended in space is contracted into pure Form. And past, present and future shrink into a single moment, which is eternity.

This amounts to saying that physics is but logic spoiled. In this proposition the whole philosophy of Ideas is summarized. And in it also is the hidden prin-

[1] We have tried to bring out what is true and what is false in this idea, so far as spatiality is concerned (see Chapter III.). It seems to us radically false as regards *duration*.

ciple of the philosophy that is innate in our understanding. If immutability is more than becoming, form is more than change, and it is by a veritable fall that the logical system of Ideas, rationally subordinated and co-ordinated among themselves, is scattered into a physical series of objects and events accidentally placed one after another. The generative idea of a poem is developed in thousands of imaginations which are materialized in phrases that spread themselves out in words. And the more we descend from the motionless idea, wound on itself, to the words that unwind it, the more room is left for contingency and choice. Other metaphors, expressed by other words, might have arisen; an image is called up by an image, a word by a word. All these words run now one after another, seeking in vain, by themselves, to give back the simplicity of the generative idea. Our ear only hears the words: it therefore perceives only accidents. But our mind, by successive bounds, leaps from the words to the images, from the images to the original idea, and so gets back, from the perception of words—accidents called up by accidents—to the conception of the Idea that posits its own being. So the philosopher proceeds, confronted with the universe. Experience makes to pass before his eyes phenomena which run, they also, one behind another in an accidental order determined by circumstances of time and place. This physical order—a degeneration of the logical order—is nothing else but the fall of the logical into space and time. But the philosopher, ascending again from the percept to the concept, sees condensed into the logical all the positive reality that the physical possesses. His intellect, doing away with the materiality that lessens being, grasps being itself in the immutable system of Ideas. Thus Science is obtained, which appears to us, complete and ready-made, as soon as we put back our intellect into its true place, correcting the deviation that separated it from the intelligible. Science is not, then, a human construction. It is prior to our intellect, independent of it, veritably the generator of Things.

And indeed, if we hold the Forms to be simply snapshots taken by the mind of the continuity of becoming, they must be relative to the mind that thinks them, they can have no independent existence. At most we might say that each of these Ideas is an *ideal*. But it is in the opposite hypothesis that we are placing ourselves. Ideas must then exist by themselves. Ancient philosophy could not escape this conclusion. Plato formulated it, and in vain did Aristotle strive to avoid it. Since movement arises from the degradation of the immutable, there could be no movement, consequently no sensible world, if there were not, somewhere, immutability realized. So, having begun by refusing to Ideas an independent existence, and finding himself nevertheless unable to deprive them of it, Aristotle pressed them into each other, rolled them up into a ball, and set above the physical world a Form that was thus found to be the Form of Forms, the Idea of Ideas, or, to use his own words, the Thought of Thought. Such is the God of Aristotle—necessarily immutable and apart from what is happening in the world, since he is only the synthesis of all concepts in a single concept. It is true that no one of the manifold concepts could exist apart, such as it is in the divine unity: in vain should we look for the ideas of Plato within the God of Aristotle. But if only we imagine the God of Aristotle in a sort of refraction of himself, or simply inclining toward the world, at once the Platonic Ideas are seen to pour themselves out of him, as if they were involved in the unity of his essence: so rays stream out from the sun, which nevertheless did not contain them. It is probably this *possibility of an outpouring* of Platonic Ideas from the Aristotelian God that is meant, in the philosophy of Aristotle, by the active intellect, the νοῦς that has been called ποιητικός—that is, by what is essential and yet unconscious in human intelligence. The νοῦς ποιητικός is Science entire, posited all at once, which the conscious, discursive intellect is condemned to reconstruct with difficulty, bit by bit. There is then within us, or rather behind us, a possible vision of God, as the Alexandrians said, a vision always virtual, never actually realized by the conscious intellect. In this intuition we should see God expand in Ideas. This it is that "does everything," [1] playing in relation to the discursive intellect, which moves in time, the same rôle as the motionless Mover himself plays in relation to the movement of the heavens and the course of things.

There is, then, immanent in the philosophy of Ideas, a particular conception of causality, which it is important to bring into full light, because it is that which each of us will reach when, in order to ascend to the origin of things, he follows to the end the natural movement of the intellect. True, the ancient philosophers never formulated it explicitly. They confined themselves to drawing the consequences of it, and, in general, they have

[1] Aristotle, *De anima*, 430 a 14 *καὶ ἔστιν ὁ μὲν τοιοῦτος νοῦς τῷ πάντα γίνεσθαι, ὁ δὲ τῷ πάντα ποιεῖν, ὡς ἕξις τὶς, οἷον τὸ φῶς. τρόπον γάρ τίνα καὶ τὸ φῶς ποιεῖ τὰ δυνάμει ὄντα χρώματα ἐνεργείᾳ χρώματα.*

marked but points of view of it rather than presented it itself. Sometimes, indeed, they speak of an *attraction*, sometimes of an *impulsion* exercised by the prime mover on the whole of the world. Both views are found in Aristotle, who shows us in the movement of the universe an aspiration of things toward the divine perfection, and consequently an ascent toward God, while he describes it elsewhere as the effect of a contact of God with the first sphere and as descending, consequently, from God to things. The Alexandrians, we think, do no more than follow this double indication when they speak of *procession* and *conversion*. Everything is derived from the first principle, and everything aspires to return to it. But these two conceptions of the divine causality can only be identified together if we bring them, both the one and the other, back to a third, which we hold to be fundamental, and which alone will enable us to understand, not only why, in what sense, things move in space and time, but also why there is space and time, why there is movement, why there are things.

This conception, which more and more shows through the reasonings of the Greek philosophers as we go from Plato to Plotinus, we may formulate thus: *The affirmation of a reality implies the simultaneous affirmation of all the degrees of reality intermediate between it and nothing*. The principle is evident in the case of number: we cannot affirm the number 10 without thereby affirming the existence of the numbers 9, 8, 7, . . ., etc.—in short, of the whole interval between 10 and zero. But here our mind passes naturally from the sphere of quantity to that of quality. It seems to us that, a certain perfection being given, the whole continuity of degradations is given also between this perfection, on the one hand, and the nought, on the other hand, that we think we conceive. Let us then posit the God of Aristotle, thought of thought—that is, thought *making a circle*, transforming itself from subject to object and from object to subject by an instantaneous, or rather an eternal, circular process: as, on the other hand, the nought appears to posit itself, and as, the two extremities being given, the interval between them is equally given, it follows that all the descending degrees of being, from the divine perfection down to the "absolute nothing," are realized automatically, so to speak, when we have posited God.

Let us then run through this interval from top to bottom. First of all, the slightest diminution of the first principle will be enough to precipitate Being into space and time; but duration and extension, which represent this first diminution, will be as near as possible to the divine inextension and eternity. We must therefore picture to ourselves this first degradation of the divine principle as a sphere turning on itself, imitating, by the perpetuity of its circular movement, the eternity of the circle of the divine thought; creating, moreover, its own place, and thereby place in general,[1] since it includes without being included and moves without stirring from the spot; creating also its own duration, and thereby duration in general, since its movement is the measure of all motion.[2] Then, by degrees, we shall see the per-

[1] *De caelo*, ii. 287 a 12 τῆς ἐσχάτης περιφορᾶς οὔτε κενόν ἐστιν ἔξωθεν οὔτε τόπος. *Phys.* iv. 212 a 34 τὸ δὲ πᾶν ἔστι μὲν ὡς κινήσεται ἔστι δ'ὡς οὔ. ὡς μὲν γὰρ ὅλον, ἅμα τὸν τόπον οὐ μεταβάλλει. κύκλῳ δὲ κινήσεται, τῶν μορίων γὰρ οὗτος ὁ τόπος.

[2] *De caelo*, i. 279 a 12 οὐδὲ χρόνος ἐστὶν ἔξω τοῦ οὐρανοῦ. *Phys.* viii. 251 b 27 ὁ χρόνος πάθος τι κινήσεως.

fection decrease, more and more, down to our sublunary world, in which the cycle of birth, growth and decay imitates and mars the original circle for the last time. So understood, the causal relation between God and the world is seen as an attraction when regarded from below, as an impulsion or a contact when regarded from above, since the first heaven, with its circular movement, is an imitation of God and all imitation is the reception of a form. Therefore, we perceive God as efficient cause or as final cause, according to the point of view. And yet neither of these two relations is the ultimate causal relation. The true relation is that which is found between the two members of an equation, when the first member is a single term and the second a sum of an endless number of terms. It is, we may say, the relation of the gold piece to the small change, if we suppose the change to offer itself automatically as soon as the gold piece is presented. Only thus can we understand why Aristotle has demonstrated the necessity of a first motionless mover, not by founding it on the assertion that the movement of things must have had a beginning, but, on the contrary, by affirming that this movement could not have begun and can never come to an end. If movement exists, or, in other words, if the small change is being counted, the gold piece is to be found somewhere. And if the counting goes on forever, having never begun, the single term that is eminently equivalent to it must be eternal. A perpetuity of mobility is possible only if it is backed by an eternity of immutability, which it unwinds in a chain without beginning or end.

Such is the last word of the Greek philosophy. We have not attempted to reconstruct it *a priori*. It has manifold origins. It is connected by many invisible threads to the soul of ancient Greece. Vain, therefore, the effort to deduce it from a simple principle.[1] But if everything that has come from poetry, religion, social life and a still rudimentary physics and biology be removed from it, if we take away all the light material that may have been used in the construction of the stately building, a solid framework remains, and this framework marks out the main lines of a metaphysic which is, we believe, the natural metaphysic of the human intellect. We come to a philosophy of this kind, indeed, whenever we follow to the end the cinematographical tendency of perception and thought. Our perception and thought begin by substituting for the continuity of evolutionary change a series of unchangeable forms which are, turn by turn, "caught on the wing," like the rings at a merry-go-round, which the children unhook with their little stick as they are passing. Now, how can the forms be passing, and on what "stick" are they strung? As the stable forms have been obtained by extracting from change everything that is definite, there is nothing left to characterize the instability on which the forms are laid, but a negative attribute, which must be indetermination itself. Such is the first proceeding of our thought: it dissociates each change into two elements—the one stable, definable for each particular case, to wit, the Form; the other indefinable and always the same, Change in general. And such, also, is the essential operation of language. Forms are all that it is capable of expressing. It is reduced to taking

[1] Especially have we left almost entirely on one side those admirable but somewhat fugitive intuitions that Plotinus was later to seize, to study and to fix.

as understood or is limited to *suggesting* a mobility which, just because it is always unexpressed, is thought to remain in all cases the same.—Then comes in a philosophy that holds the dissociation thus effected by thought and language to be legitimate. What can it do, except objectify the distinction with more force, push it to its extreme consequences, reduce it into a system? It will therefore construct the real, on the one hand, with definite Forms or immutable elements, and, on the other, with a principle of mobility which, being the negation of the form, will, by the hypothesis, escape all definition and be the purely indeterminate. The more it directs its attention to the forms delineated by thought and expressed by language, the more it will see them rise above the sensible and become subtilized into pure concepts, capable of entering one within the other, and even of being at last massed together into a single concept, the synthesis of all reality, the achievement of all perfection. The more, on the contrary, it descends toward the invisible source of the universal mobility, the more it will feel this mobility sink beneath it and at the same time become void, vanish into what it will call the "non-being." Finally, it will have on the one hand the system of ideas, logically co-ordinated together or concentrated into one only, on the other a quasi-nought, the Platonic "non-being" or the Aristotelian "matter."—But, having cut your cloth, you must sew it. With supra-sensible Ideas and an infra-sensible non-being, you now have to reconstruct the sensible world. You can do so only if you postulate a kind of metaphysical necessity in virtue of which the confronting of this All with this Zero *is equivalent* to the affirmation of all the degrees of reality that measure the interval between them—just as an undivided number, when regarded as a difference between itself and zero, is revealed as a certain sum of units, and with its own affirmation affirms all the lower numbers. That is the natural postulate. It is that also that we perceive as the base of the Greek philosophy. In order then to explain the specific characters of each of these degrees of intermediate reality, nothing more is necessary than to measure the distance that separates it from the integral reality. Each lower degree consists in a diminution of the higher, and the *sensible* newness that we perceive in it is resolved, from the point of view of the *intelligible*, into a new quantity of negation which is superadded to it. The smallest possible quantity of negation, that which is found already in the highest forms of sensible reality, and consequently *a fortiori* in the lower forms, is that which is expressed by the most general attributes of sensible reality, extension and duration. By increasing degradations we will obtain attributes more and more special. Here the philosopher's fancy will have free scope, for it is by an arbitrary decree, or at least a debatable one, that a particular aspect of the sensible world will be equated with a particular diminution of being. We shall not necessarily end, as Aristotle did, in a world consisting of concentric spheres turning on themselves. But we shall be led to an analogous cosmology—I mean, to a construction whose pieces, though all different, will have none the less the same relations between them. And this cosmology will be ruled by the same principle. The physical will be defined by the logical. Beneath the changing phenomena will appear to us, by transparence, a closed system of concepts subordinated to and co-ordinated with each other. Science, understood as the system of concepts, will be more real than the sensible reality. It will be prior to human knowledge, which is only able to spell it letter by letter; prior also to things, which awkwardly try to imitate it. It would only have to be diverted an instant from itself in order to step out of its eternity and thereby coincide with all this knowledge and all these things. Its immutability is therefore, indeed, the cause of the universal becoming.

Such was the point of view of ancient philosophy in regard to change and duration. That modern philosophy has repeatedly, but especially in its beginnings, had the wish to depart from it, seems to us unquestionable. But an irresistible attraction brings the intellect back to its natural movement, and the metaphysic of the moderns to the general conclusions of the Greek metaphysic. We must try to make this point clear, in order to show by what invisible threads our mechanistic philosophy remains bound to the ancient philosophy of Ideas, and how also it responds to the requirements, above all practical, of our understanding.

Modern, like ancient, science proceeds according to the cinematographical method. It cannot do otherwise; all science is subject to this law. For it is of the essence of science to handle *signs*, which it substitutes for the objects themselves. These signs undoubtedly differ from those of language by their greater precision and their higher efficacy; they are none the less tied down to the general condition of the sign, which is to denote a fixed aspect of the reality under an arrested form. In order to think movement, a constantly renewed effort of the mind is necessary. Signs are made to dispense us with this effort by substituting, for the moving continuity of things, an artificial reconstruction which is its equivalent in practice and has the advantage of being easily handled. But let us leave aside the means and consider only the end. What is the essential object of science? It is to enlarge our influence over things. Science may be speculative in its form, disinterested in its immediate ends: in other words we may give it as long a credit as it wants. But, however long the day of reckoning may be put off, some time or other the payment must be made. It is always then, in short, practical utility that science has in view. Even when it launches into theory, it is bound to adapt its behavior to the general form of practice. However high it may rise, it must be ready to fall back into the field of action, and at once to get on its feet. This would not be possible for it, if its rhythm differed absolutely from that of action itself. Now action, we have said, proceeds by leaps. To act is to re-adapt oneself. To know, that is to say, to foresee in order to act, is then to go from situation to situation, from arrangement to rearrangement. Science may consider rearrangements that come closer and closer to each other; it may thus increase the number of moments that it isolates, but it always isolates moments. As to what happens in the interval between the moments, science is no more concerned with that than are our common intelligence, our senses and our language: it does not bear on the interval, but only on the extremities. So the cinematographical method forces itself upon our science, as it did already on that of the ancients.

Wherein, then, is the difference between the two

sciences? We indicated it when we said that the ancients reduced the physical order to the vital order, that is to say, laws to genera, while the moderns try to resolve genera into laws. But we have to look at it in another aspect, which, moreover, is only a transposition, of the first. Wherein consists the difference of attitude of the two sciences toward change? We may formulate it by saying that *ancient science thinks it knows its object sufficiently when it has noted of it some privileged moments, whereas modern science considers the object at any moment whatever.*

The forms or ideas of Plato or of Aristotle correspond to privileged or salient moments in the history of things —those, in general, that have been fixed by language. They are supposed, like the childhood or the old age of a living being, to characterize a period of which they express the quintessence, all the rest of this period being filled by the passage, of no interest in itself, from one form to another form. Take, for instance, a falling body. It was thought that we got near enough to the fact when we characterized it as a whole: it was a movement *downward*; it was the tendency toward a *center*; it was the *natural* movement of a body which, separated from the earth to which it belonged, was now going to find its place again. They noted, then, the final term or culminating point (τέλος, ἀκμή) and set it up as the essential moment: this moment, that language has retained in order to express the whole of the fact, sufficed also for science to characterize it. In the physics of Aristotle, it is by the concepts "high" and "low," spontaneous displacement and forced displacement, own place and strange place, that the movement of a body shot into space or falling freely is defined. But Galileo thought there was no essential moment, no privileged instant. To study the falling body is to consider it at it matters not what moment in its course. The true science of gravity is that which will determine, for any moment of time whatever, the position of the body in space. For this, indeed, signs far more precise than those of language are required.

We may say, then, that our physics differs from that of the ancients chiefly in the indefinite breaking up of time. For the ancients, time comprised as many undivided periods as our natural perception and our language cut out in it successive facts, each presenting a kind of individuality. For that reason, each of these facts admits, in their view, of only a *total* definition or description. If, in describing it, we are led to distinguish phases in it, we have several facts instead of a single one, several undivided periods instead of a single period; but time is always supposed to be divided into determinate periods, and the mode of division to be forced on the mind by apparent crises of the real, comparable to that of puberty, by the apparent release of a new form.—For a Kepler, or a Galileo, on the contrary, time is not divided objectively in one way or another by the matter that fills it. It has no natural articulations. We can, we ought to, divide it as we please. All moments count. None of them has the right to set itself up as a moment that represents or dominates the others. And, consequently, we know a change only when we are able to determine what it is about at any one of its moments.

The difference is profound. In fact, in a certain aspect it is radical. But, from the point of view from which we are regarding it, it is a difference of degree rather than of kind. The human mind has passed from the first kind of knowledge to the second through gradual perfecting, simply by seeking a higher precision. There is the same relation between these two sciences as between the noting of the phases of a movement by the eye and the much more complete recording of these phases by instantaneous photography. It is the same cinematographical mechanism in both cases, but it reaches a precision in the second that it cannot have in the first. Of the gallop of a horse our eye perceives chiefly a characteristic, essential or rather schematic attitude, a form that appears to radiate over a whole period and so fill up a time of gallop. It is this attitude that sculpture has fixed on the frieze of the Parthenon. But instantaneous photography isolates any moment; it puts them all in the same rank, and thus the gallop of a horse spreads out for it into as many successive attitudes as it wishes, instead of massing itself into a single attitude, which is supposed to flash out in a privileged moment and to illuminate a whole period.

From this original difference flow all the others. A science that considers, one after the other, undivided periods of duration, sees nothing but phases succeeding phases, forms replacing forms; it is content with a *qualitative* description of objects, which it likens to organized beings. But when we seek to know what happens within one of these periods, at any moment of time, we are aiming at something entirely different. The changes which are produced from one moment to another are no longer, by the hypothesis, changes of quality; they are *quantitative* variations, it may be of the phenomenon itself, it may be of its elementary parts. We were right then to say that modern science is distinguishable from the ancient in that it applies to magnitudes and proposes first and foremost to measure them. The ancients did indeed try experiments, and on the other hand Kepler tried no experiment, in the proper sense of the word, in order to discover a law which is the very type of scientific knowledge as we understand it. What distinguishes modern science is not that it is experimental, but that it experiments and, more generally, works only with a view to measure.

For that reason it is right, again, to say that ancient science applied to *concepts*, while modern science seeks *laws*—constant relations between variable magnitudes. The concept of circularity was sufficient to Aristotle to define the movement of the heavenly bodies. But, even with the more accurate concept of elliptical form, Kepler did not think he had accounted for the movement of planets. He had to get a law, that is to say, a constant relation between the quantitative variations of two or several elements of the planetary movement.

Yet these are only consequences—differences that follow from the fundamental difference. It did happen to the ancients accidentally to experiment with a view to measuring, as also to discover a law expressing a constant relation between magnitudes. The principle of Archimedes is a true experimental law. It takes into account three variable magnitudes: the volume of a body, the density of the liquid in which the body is immersed, the vertical pressure that is being exerted. And it states indeed that one of these three terms is a function of the other two.

The essential, original difference must therefore be sought elsewhere. It is the same that we noticed first. The science of the ancients is static. Either it considers in block the change that it studies, or, if it divides the change into periods, it makes of each of these periods a block in its turn: which amounts to saying that it takes no account of time. But modern science has been built up around the discoveries of Galileo and of Kepler, which immediately furnished it with a model. Now, what do the laws of Kepler say? They lay down a relation between the areas described by the heliocentric radius-vector of a planet and the *time* employed in describing them, a relation between the longer axis of the orbit and the *time* taken up by the course. And what was the principle discovered by Galileo? A law which connected the space traversed by a falling body with the *time* occupied by the fall. Furthermore, in what did the first of the great transformations of geometry in modern times consist, if not in introducing—in a veiled form, it is true—time and movement even in the consideration of figures? For the ancients, geometry was a purely static science. Figures were given to it at once, completely finished, like the Platonic Ideas. But the essence of the Cartesian geometry (although Descartes did not give it this form) was to regard every plane curve as described by the movement of a point on a movable straight line which is displaced, parallel to itself, along the axis of the abscissae—the displacement of the movable straight line being supposed to be uniform and the abscissa thus becoming representative of the time. The curve is then defined if we can state the relation connecting the space traversed on the movable straight line to the time employed in traversing it, that is, if we are able to indicate the position of the movable point, on the straight line which it traverses, at any moment whatever of its course. This relation is just what we call the equation of the curve. To substitute an equation for a figure consists, therefore, in seeing the actual position of the moving points in the tracing of the curve at any moment whatever, instead of regarding this tracing all at once, gathered up in the unique moment when the curve has reached its finished state.

Such, then, was the directing idea of the reform by which both the science of nature and mathematics, which serves as its instrument, were renewed. Modern science is the daughter of astronomy; it has come down from heaven to earth along the inclined plane of Galileo, for it is through Galileo that Newton and his successors are connected with Kepler. Now, how did the astronomical problem present itself to Kepler? The question was, knowing the respective positions of the planets at a given moment, how to calculate their positions at any other moment. So the same question presented itself, henceforth, for every material system. Each material point became a rudimentary planet, and the main question, the ideal problem whose solution would yield the key to all the others was, the positions of these elements at a particular moment being given, how to determine their relative positions at any moment. No doubt the problem cannot be put in these precise terms except in very simple cases, for a schematized reality; for we never know the respective positions of the real elements of matter, supposing there are real elements; and, even if we knew them at a given moment, the calculation of their positions at another moment would generally require a mathematical effort surpassing human powers. But it is enough for us to know that these elements might be known, that their present positions might be noted, and that a superhuman intellect might, by submitting these data to mathematical operations, determine the positions of the elements at any other moment of time. This conviction is at the bottom of the questions we put to ourselves on the subject of nature, and of the methods we employ to solve them. That is why every law in static form seems to us as a provisional instalment or as a particular view of a dynamic law which alone would give us whole and definitive knowledge.

Let us conclude, then, that our science is not only distinguished from ancient science in this, that it seeks laws, nor even in this, that its laws set forth relations between magnitudes: we must add that the magnitude to which we wish to be able to relate all others is time, and that *modern science must be defined pre-eminently by its aspiration to take time as an independent variable.* But with what time has it to do?

We have said before, and we cannot repeat too often, that the science of matter proceeds like ordinary knowledge. It perfects this knowledge, increases its precision and its scope, but it works in the same direction and puts the same mechanism into play. If, therefore, ordinary knowledge, by reason of the cinematographical mechanism to which it is subjected, forbears to follow becoming in so far as becoming is moving, the science of matter renounces it equally. No doubt, it distinguishes as great a number of moments as we wish in the interval of time it considers. However small the intervals may be at which it stops, it authorizes us to divide them again if necessary. In contrast with ancient science, which stopped at certain so-called essential moments, it is occupied indifferently with any moment whatever. But it always considers moments, always virtual stopping-places, always, in short, immobilities. Which amounts to saying that real time, regarded as a flux, or, in other words, as the very mobility of being, escapes the hold of scientific knowledge. We have already tried to establish this point in a former work. We alluded to it again in the first chapter of this book. But it is necessary to revert to it once more, in order to clear up misunderstandings.

When positive science speaks of time, what it refers to is the movement of a certain mobile T on its trajectory. This movement has been chosen by it as representative of time, and it is, by definition, uniform. Let us call T_1, T_2, T_3, . . . etc., points which divide the trajectory of the mobile into equal parts from its origin T_0. We shall say that 1, 2, 3, . . . units of time have flowed past, when the mobile is at the points T_1, T_2, T_3, . . . of the line it traverses. Accordingly, to consider the state of the universe at the end of a certain time *t*, is to examine where it will be when T is at the point T_t of its course. But of the *flux* itself of time, still less of its effect on consciousness, there is here no question; for there enter into the calculation only the points T_1, T_2, T_3, . . . taken on the flux, never the flux itself. We may narrow the time considered as much as we will, that is, break up at will the interval between two consecutive divisions T_n and T_{n+1}; but it is always with points, and with points only, that we are dealing. What we retain of the movement of the mobile T are positions taken on its trajectory. What we retain of all the other points of the universe are their positions on their respective trajectories. To each *virtual stop* of the moving body T at the points of division T_1, T_2, T_3, . . . we make correspond a *virtual stop* of all the other mobiles at the points where they are passing. And when we say that a movement or any other change has occupied a time *t*, we mean by it that we have noted a number *t* of correspondences of this kind. We have therefore counted simultaneities; we have not concerned ourselves with the flux that goes from one to another. The proof of this is that I can, at discretion, vary the rapidity of the flux of the universe in regard to a consciousness that is independent of it and that would perceive the variation by the quite qualitative *feeling* that it would have of it: whatever the variation had been, since the movement of T would participate in this variation, I should have nothing to change in my equations nor in the numbers that figure in them.

Let us go further. Suppose that the rapidity of the flux becomes infinite. Imagine, as we said in the first pages of this book, that the trajectory of the mobile T is given at once, and that the whole history, past, present and future, of the material universe is spread out instantaneously in space. The same mathematical correspondences will subsist between the moments of the history of the world unfolded like a fan, so to speak, and the divisions T_1, T_2, T_3, . . . of the line which will be called, by definition, "the course of time." In the eyes of science nothing will have changed. But if, time thus spreading itself out in space and succession becoming juxtaposition, science has nothing to change in what it tells us, we must conclude that, in what it tells us, it takes account neither of *succession* in what of it is specific nor of *time* in what there is in it that is fluent. It has no sign to express what strikes our consciousness in succession and duration. It no more applies to becoming, so far as that is moving, than the bridges thrown here and there across the stream follow the water that flows under their arches.

Yet succession exists; I am conscious of it; it is a fact. When a physical process is going on before my eyes, my perception and my inclination have nothing to do with accelerating or retarding it. What is important to the physicist is the *number* of units of duration the process fills; he does not concern himself about the units themselves and that is why the successive states of the world might be spread out all at once in space without his having to change anything in his science or to cease talking about time. But for us, conscious beings, it is the units that matter, for we do not count extremities of intervals, we feel and live the intervals themselves. Now, we are conscious of these intervals as of *definite* intervals. Let me come back again to the sugar in my glass of water:[1] why must I wait for it to melt? While the duration of the phenomenon is *relative* for the physicist, since it is reduced to a certain number of units of time and the units themselves are indifferent, this duration is an *absolute* for my consciousness, for it coincides with a certain degree of impatience which is rigorously determined. Whence comes this determination? What is it that obliges me to wait, and to wait for a certain length of psychical duration which is forced upon me, over which I have no power? If succession, in so far as distinct from mere juxtaposition, has no real efficacy, if time is not a

[1] See page 12.

kind of force, why does the universe unfold its successive states with a velocity which, in regard to my consciousness, is a veritable absolute? Why with this particular velocity rather than any other? Why not with an infinite velocity? Why, in other words, is not everything given at once, as on the film of the cinematograph? The more I consider this point, the more it seems to me that, if the future is bound to *succeed* the present instead of being given alongside of it, it is because the future is not altogether determined at the present moment, and that if the time taken up by this succession is something other than a number, if it has for the consciousness that is installed in it absolute value and reality, it is because there is unceasingly being created in it, not indeed in any such artificially isolated system as a glass of sugared water, but in the concrete whole of which every such system forms part, something unforeseeable and new. This duration may not be the fact of matter itself, but that of the life which reascends the course of matter; the two movements are none the less mutually dependent upon each other. *The duration of the universe must therefore be one with the latitude of creation which can find place in it.*

When a child plays at reconstructing a picture by putting together the separate pieces in a puzzle game, the more he practices, the more and more quickly he succeeds. The reconstruction was, moreover, instantaneous, the child found it ready-made, when he opened the box on leaving the shop. The operation, therefore, does not require a definite time, and indeed, theoretically, it does not require any time. That is because the result is given. It is because the picture is already created, and because to obtain it requires only a work of recomposing and rearranging—a work that can be supposed going faster and faster, and even infinitely fast, up to the point of being instantaneous. But, to the artist who creates a picture by drawing it from the depths of his soul, time is no longer an accessory; it is not an interval that may be lengthened or shortened without the content being altered. The duration of his work is part and parcel of his work. To contract or to dilate it would be to modify both the psychical evolution that fills it and the invention which is its goal. The time taken up by the invention is one with the invention itself. It is the progress of a thought which is changing in the degree and measure that it is taking form. It is a vital process, something like the ripening of an idea.

The painter is before his canvas, the colors are on the palette, the model is sitting—all this we see, and also we know the painter's style: do we foresee what will appear on the canvas? We possess the elements of the problem; we know in an abstract way, how it will be solved, for the portrait will surely resemble the model and will surely resemble also the artist; but the concrete solution brings with it that unforeseeable nothing which is everything in a work of art. And it is this nothing that takes time. Nought as matter, it creates itself as form. The sprouting and flowering of this form are stretched out on an unshrinkable duration, which is one with their essence. So of the works of nature. Their novelty arises from an internal impetus which is progress or succession, which confers on succession a peculiar virtue or which owes to succession the whole of its virtue—which, at any rate, makes succession, or *continuity of interpenetration* in time, irreducible to a mere instantaneous juxtaposition in space. This is why the idea of reading in a present state of the material universe the future of living forms, and of unfolding now their history yet to come, involves a veritable absurdity. But this absurdity is difficult to bring out, because our memory is accustomed to place alongside of each other, in an ideal space, the terms it perceives in turn, because it always represents *past* succession in the form of juxtaposition. It is able to do so, indeed, just because the past belongs to that which is already invented, to the dead, and no longer to creation and to life. Then, as the succession to come will end by being a succession past, we persuade ourselves that the duration to come admits of the same treatment as past duration, that it is, even now, unrollable, that the future is there, rolled up, already painted on the canvas. An illusion, no doubt, but an illusion that is natural, ineradicable, and that will last as long as the human mind!

Time is invention or it is nothing at all. But of time-invention physics can take no account, restricted as it is to the cinematographical method. It is limited to counting simultaneities between the events that make up this time and the positions of the mobile T on its trajectory. It detaches these events from the whole, which at every moment puts on a new form and which communicates to them something of its novelty. It considers them in the abstract, such as they would be outside of the living whole, that is to say, in a time unrolled in space. It retains only the events or systems of events that can be thus isolated without being made to undergo too profound a deformation, because only these lend themselves to the application of its method. Our physics dates from the day when it was known how to isolate such systems. To sum up, *while modern physics is distinguished from ancient physics by the fact that it considers any moment of time whatever, it rests altogether on a substitution of time-length for time-invention.*

It seems then that, parallel to this physics, a second kind of knowledge ought to have grown up, which could have retained what physics allowed to escape. On the flux itself of duration science neither would nor could lay hold, bound as it was to the cinematographical method. This second kind of knowledge would have set the cinematographical method aside. It would have called upon the mind to renounce its most cherished habits. It is within becoming that it would have transported us by an effort of sympathy. We should no longer be asking where a moving body will be, what shape a system will take, through what state a change will pass at a given moment: the moments of time, which are only arrests of our attention, would no longer exist; it is the flow of time, it is the very flux of the real that we should be trying to follow. The first kind of knowledge has the advantage of enabling us to foresee the future and of making us in some measure masters of events; in return, it retains of the moving reality only eventual immobilities, that is to say, views taken of it by our mind. It symbolizes the real and transposes it into the human rather than expresses it. The other knowledge, if it is possible, is practically useless, it will not extend our empire over nature, it will even go against certain natural aspirations of the intellect; but, if it succeeds, it is reality itself that it will hold in a firm and final embrace. Not only may we thus complete the intellect and its knowledge of matter by accustoming it to install itself within the moving, but by developing also another faculty, complementary to the intellect, we may open a perspective on the other half of the real. For, as soon as we are confronted with true duration, we see that it means creation, and that if that which is being unmade endures, it can only be because it is inseparably bound to what is making itself. Thus will appear the necessity of a continual growth of the universe, I should say of a *life* of the real. And thus will be seen in a new light the life which we find on the surface of our planet, a life directed the same way as that of the universe, and inverse of materiality. To intellect, in short, there will be added intuition.

The more we reflect on it, the more we shall find that this conception of metaphysics is that which modern science suggests.

For the ancients, indeed, time is theoretically negligible, because the duration of a thing only manifests the degradation of its essence: it is with this motionless essence that science has to deal. Change being only the effort of a form toward its own realization, the realization is all that it concerns us to know. No doubt the realization is never complete: it is this that ancient philosophy expresses by saying that we do not perceive form without matter. But if we consider the changing object at a certain essential moment, at its apogee, we may say that there it just touches its intelligible form. This intelligible form, this ideal and, so to speak, limiting form, our science seizes upon. And possessing in this the gold-piece, it holds eminently the small money which we call becoming or change. This change is less than being. The knowledge that would take it for object, supposing such knowledge were possible, would be less than science.

But, for a science that places all the moments of time in the same rank, that admits no essential moment, no culminating point, no apogee, change is no longer a diminution of essence, duration is not a dilution of eternity. The flux of time is the reality itself, and the things which we study are the things which flow. It is true that of this flowing reality we are limited to taking instantaneous views. But, just because of this, scientific knowledge must appeal to another knowledge to complete it. While the ancient conception of scientific knowledge ended in making time a degradation, and change the diminution of a form given from all eternity—on the contrary, by following the new conception to the end, we should come to see in time a progressive growth of the absolute, and in the evolution of things a continual invention of forms ever new.

It is true that it would be to break with the metaphysics of the ancients. They saw only one way of knowing definitely. Their science consisted in a scattered and fragmentary metaphysics, their metaphysics in a concentrated and systematic science. Their science and metaphysics were, at most, two species of one and the same genus. In our hypothesis, on the contrary, science and metaphysics are two opposed although complementary ways of knowing, the first retaining only moments, that is to say, that which does not endure, the second bearing on duration itself. Now, it was natural to hesitate between so novel a conception of metaphysics and the traditional conception. The temptation must have been strong to repeat with the new science what had been tried on the old, to suppose our scientific knowledge of nature completed at once, to unify it entirely, and to give to this unification, as the Greeks had already done, the name of metaphysics. So, beside the new way that philosophy might have prepared, the old remained open, that indeed which physics trod. And, as physics retained of time only what could as well be spread out all at once in space, the metaphysics that chose the same direction had necessarily to proceed as if time created and annihilated nothing, as if duration had no efficacy. Bound, like the physics of the moderns and the metaphysics of the ancients, to the cinematographical method, it ended with the conclusion, implicitly admitted at the start and immanent in the method itself: *All is given.*

That metaphysics hesitated at first between the two paths seems to us unquestionable. The indecision is visible in Cartesianism. On the one hand, Descartes affirms universal mechanism: from this point of view movement would be relative,[1] and, as time has just as much reality as movement, it would follow that past, present and future are given from all eternity. But, on the other hand (and that is why the philosopher has not gone to these extreme consequences), Descartes believes in the free will of man. He superposes on the determinism of physical phenomena the indeterminism of human actions, and, consequently, on time-length a time in which there is invention, creation, true succession. This duration he supports on a God who is unceasingly renewing the creative act, and who, being thus tangent to time and becoming, sustains them, communicates to them necessarily some-

[1] Descartes. *Principes*, ii. § 29.

thing of his absolute reality. When he places himself at this second point of view, Descartes speaks of movement, even spatial, as of an absolute.[1]

He therefore entered both roads one after the other, having resolved to follow neither of them to the end. The first would have led him to the denial of free will in man and of real will in God. It was the suppression of all efficient duration, the likening of the universe to a thing *given*, which a superhuman intelligence would embrace at once in a moment or in eternity. In following the second, on the contrary, he would have been led to all the consequences which the intuition of true duration implies. Creation would have appeared not simply as *continued*, but also as *continuous*. The universe, regarded as a whole, would really evolve. The future would no longer be determinable by the present; at most we might say that, once realized, it can be found again in its antecedents, as the sounds of a new language can be expressed with the letters of an old alphabet if we agree to enlarge the value of the letters and to attribute to them, retroactively, sounds which no combination of the old sounds could have produced beforehand. Finally, the mechanistic explanation might have remained universal in this, that it can indeed be extended to as many systems as we choose to cut out in the continuity of the universe; but mechanism would then have become a *method* rather than a *doctrine*. It would have expressed the fact that science must proceed after the cinematographical manner, that the function of science is to scan the rhythm of the flow of things and not to fit itself into that flow.—

[1] Descartes, *Principes*, ii. §§ 36 ff.

Such were the two opposite conceptions of metaphysics which were offered to philosophy.

It chose the first. The reason of this choice is undoubtedly the mind's tendency to follow the cinematographical method, a method so natural to our intellect, and so well adjusted also to the requirements of our science, that we must feel doubly sure of its speculative impotence to renounce it in metaphysics. But ancient philosophy also influenced the choice. Artists forever admirable, the Greeks created a type of suprasensible truth, as of sensible beauty, whose attraction is hard to resist. As soon as we incline to make metaphysics a systematization of science, we glide in the direction of Plato and of Aristotle. And, once in the zone of attraction in which the Greek philosophers moved, we are drawn along in their orbit.

Such was the case with Leibniz, as also with Spinoza. We are not blind to the treasures of originality their doctrines contain. Spinoza and Leibniz have poured into them the whole content of their souls, rich with the inventions of their genius and the acquisitions of modern thought. And there are in each of them, especially in Spinoza, flashes of intuition that break through the system. But if we leave out of the two doctrines what breathes life into them, if we retain the skeleton only, we have before us the very picture of Platonism and Aristotelianism seen through Cartesian mechanism. They present to us a systematization of the new physics, constructed on the model of the ancient metaphysics.

What, indeed, could the unification of physics be? The inspiring idea of that science was to isolate, within the universe, systems of material points such that, the position of each of these points being known at a given moment, we could then calculate it for any moment whatever. As, moreover, the systems thus defined were the only ones on which the new science had hold, and as it could not be known beforehand whether a system satisfied or did not satisfy the desired condition, it was useful to proceed always and everywhere *as if* the condition was realized. There was in this a methodological rule, a very natural rule—so natural, indeed, that it was not even necessary to formulate it. For simple common sense tells us that when we are possessed of an effective instrument of research, and are ignorant of the limits of its applicability, we should act as if its applicability were unlimited; there will always be time to abate it. But the temptation must have been great for the philosopher to hypostatize this hope, or rather this impetus, of the new science, and to convert a general rule of method into a fundamental law of things. So he transported himself at once to the limit; he supposed physics to have become complete and to embrace the whole of the sensible world. The universe became a system of points, the position of which was rigorously determined at each instant by relation to the preceding instant and theoretically calculable for any moment whatever. The result, in short, was universal mechanism. But it was not enough to formulate this mechanism; what was required was to found it, to give the reason for it and prove its necessity. And the essential affirmation of mechanism being that of a reciprocal mathematical dependence of all the points of the universe, as also of all the moments of the universe, the reason of mechanism had to be discovered in the unity of a principle into which could be contracted all that is juxtaposed in space and successive in time. Hence, the whole of the real was supposed to be given at once. The reciprocal determination of the juxtaposed appearances in space was explained by the indivisibility of true being, and the inflexible determinism of successive phenomena in time simply expressed that the whole of being is given in the eternal.

The new philosophy was going, then, to be a recommencement, or rather a transposition, of the old. The ancient philosophy had taken each of the *concepts* into which a becoming is concentrated or which mark its apogee: it supposed them all known, and gathered them up into a single concept, form of forms, idea of ideas, like the God of Aristotle. The new philosophy was going to take each of the *laws* which condition a becoming in relation to others and which are as the permanent substrata of phenomena: it would suppose them all known, and would gather them up into a unity which also would express them eminently, but which, like the God of Aristotle and for the same reasons, must remain immutably shut up in itself.

True, this return to the ancient philosophy was not without great difficulties. When a Plato, an Aristotle, or a Plotinus melt all the concepts of their science into a single one, in so doing they embrace the whole of the real, for concepts are supposed to represent the things themselves, and to possess at least as much positive content. But a law, in general, expresses only a relation, and physical laws in particular express only *quantitative* relations between concrete things. So that if a modern philosopher works with the laws of the new science as the Greek philosopher did with the concepts of the ancient science, if he makes all the conclusions of a physics supposed omniscient converge on a single point, he neglects what is concrete in the phenomena—the qualities perceived, the perceptions themselves. His synthesis comprises, it seems, only a fraction of reality. In fact, the first result of the new science was to cut the real into two halves, quantity and quality, the former being credited to the account of *bodies* and the latter to the account of *souls.* The ancients had raised no such barriers either between quality and quantity or between soul and body. For them, the mathematical concepts were concepts like the others, related to the others and fitting quite naturally into the hierarchy of the Ideas. Neither was the body then defined by geometrical extension, nor the soul by consciousness. If the ψυχή of Aristotle, the entelechy of a living body, is less spiritual than our "soul," it is because his σῶμα, already impregnated with the Idea, is less corporeal than our "body." The scission was not yet irremediable between the two terms. It has become so, and thence a metaphysic that aims at an abstract unity must resign itself either to comprehend in its synthesis only one half of the real, or to take advantage of the absolute heterogeneity of the two halves in order to consider one as a translation of the other. Different phrases will express different things if they belong to the same language, that is to say, if there is a certain relationship of sound between them. But if they belong to two different languages, they might, just because of their radical diversity of sound, express the same thing. So of quality and quantity, of soul and body. It is for having cut all connection between the two terms that philosophers have been led to establish between them a rigorous parallelism, of which the ancients had not dreamed, to regard them as translations and not as inversions of each other; in short, to posit a fundamental identity as a substratum to their duality. The synthesis to which they rose thus became capable of embracing everything. A divine mechanism made the phenomena of thought to correspond to those of extension, each to each, qualities to quantities, souls to bodies.

It is this parallelism that we find both in Leibniz and in Spinoza—in different forms, it is true, because of the unequal importance which they attach to extension. With Spinoza, the two terms Thought and Extension are placed, in principle at least, in the same rank. They are, therefore, two translations of one and the same original, or, as Spinoza says, two attributes of one and the same substance, which we must call God. And these two translations, as also an infinity of others into languages which we know not, are called up and even forced into existence by the original, just as the essence of the circle is translated automatically, so to speak, both by a figure and by an equation. For Leibniz, on the contrary, extension is indeed still a translation, but it is thought that is the original, and thought might dispense with translation, the translation being made only for us. In positing God, we necessarily posit also all the possible views of God, that is to say, the monads. But we can always imagine that a view has been taken from a point of view, and it is natural for an imperfect mind like ours to class views, qualitatively different, according to the order and position of points of view, qualitatively identical, from which the views might have been taken. In reality the points of view do not exist, for there are only views, each given in an indivisible block and representing in its own way the whole of reality, which is God. But we need to express the plurality of the views, that are *unlike* each other, by the multiplicity of the points of view that are *exterior* to each other; and we also need to symbolize the more or less close relationship between the views by the relative situation of the points of view to one another, their nearness or their distance, that is to say, by a magnitude. That is what Leibniz means when he says that space is the order of coexistents, that the perception of extension is a confused perception (that is to say, a perception relative to an imperfect mind), and that nothing exists but monads, expressing thereby that the real Whole has no parts, but is repeated to infinity, each time integrally (though diversely) within itself, and that all these repetitions are complementary to each other. In just the same way, the visible relief of an object is equivalent to the whole set of stereoscopic views taken of it from all points, so that, instead of seeing in the relief a juxtaposition of solid parts, we might quite as well look upon it as made of the *reciprocal complementarity* of these whole views, each given in block, each indivisible, each different from all the others and yet representative of the same thing. The Whole, that is to say, God, is this very relief for Leibniz, and the monads are these complementary plane views; for that reason he defines God as "the substance that has no point of view," or, again, as "the universal harmony," that is to say, the reciprocal complementarity of monads. In short, Leibniz differs from Spinoza in this, that he looks upon the universal mechanism as an aspect which reality takes for us, whereas,

Spinoza makes of it an aspect which reality takes for itself.

It is true that, after having concentrated in God the whole of the real, it became difficult for them to pass from God to things, from eternity to time. The difficulty was even much greater for these philosophers than an Aristotle or a Plotinus. The God of Aristotle, indeed, had been obtained by the compression and reciprocal compenetration of the Ideas that represent, in their finished state or in their culminating point, the changing things of the world. He was, therefore, transcendent to the world, and the duration of things was juxtaposed to His eternity, of which it was only a weakening. But in the principle to which we are led by the consideration of universal mechanism, and which must serve as its substratum, it is not concepts or *things*, but laws or *relations* that are condensed. Now, a relation does not exist separately. A law connects changing terms and is immanent in what it governs. The principle in which all these relations are ultimately summed up, and which is the basis of the unity of nature, cannot, therefore, be transcendent to sensible reality; it is immanent in it, and we must suppose that it is at once both in and out of time, gathered up in the unity of its substance and yet condemned to wind it off in an endless chain. Rather than formulate so appalling a contradiction, the philosophers were necessarily led to sacrifice the weaker of the two terms, and to regard the temporal aspect of things as a mere illusion. Leibniz says so in explicit terms, for he makes of time, as of space, a confused perception. While the multiplicity of his monads expresses only the diversity of views taken of the whole, the history of an isolated monad seems to be hardly anything else than the manifold views that it can take of its own substance: so that time would consist in all the points of view that each monad can assume toward itself, as space consists in all the points of view that all monads can assume toward God. But the thought of Spinoza is much less clear, and this philosopher seems to have sought to establish, between eternity and that which has duration, the same difference as Aristotle made between essence and accidents: a most difficult undertaking, for the υλη of Aristotle was no longer there to measure the distance and explain the passage from the essential to the accidental, Descartes having eliminated it forever. However that may be, the deeper we go into the Spinozistic conception of the "inadequate," as related to the "adequate," the more we feel ourselves moving in the direction of Aristotelianism —just as the Leibnizian monads, in proportion as they mark themselves out the more clearly, tend to approximate to the Intelligibles of Plotinus.[1] The natural trend of these two philosophies brings them back to the conclusions of the ancient philosophy.

To sum up, the resemblances of this new metaphysic to that of the ancients arise from the fact that both suppose ready-made—the former above the sensible, the latter within the sensible—a science one and complete, with which any reality that the sensible may contain is believed to coincide. *For both, reality as well as truth are*

[1] In a course of lectures on Plotinus, given at the Collège de France in 1897-1898, we tried to bring out these resemblances. They are numerous and impressive. The analogy is continued even in the formulae employed on each side.

integrally given in eternity. Both are opposed to the idea of a reality that creates itself gradually, that is, at bottom, to an absolute duration.

Now, it might easily be shown that the conclusions of this metaphysic, springing from science, have rebounded upon science itself, as it were, by ricochet. They penetrate the whole of our so-called empiricism. Physics and chemistry study only inert matter; biology, when it treats the living being physically and chemically, considers only the inert side of the living: hence the mechanistic explanations, in spite of their development, include only a small part of the real. To suppose *a priori* that the whole of the real is resolvable into elements of this kind, or at least that mechanism can give a complete translation of what happens in the world, is to pronounce for a certain metaphysic—the very metaphysic of which Spinoza and Leibniz have laid down the principles and drawn the consequences. Certainly, the psycho-physiologist who affirms the exact equivalence of the cerebral and the psychical state, who imagines the possibility, for some superhuman intellect, of reading in the brain what is going on in consciousness, believes himself very far from the metaphysicians of the seventeenth century, and very near to experience. Yet experience pure and simple tells us nothing of the kind. It shows us the interdependence of the mental and the physical, the necessity of a certain cerebral substratum for the psychical state —nothing more. From the fact that two things are mutually dependent, it does not follow that they are equivalent. Because a certain screw is necessary to a certain machine, because the machine works when the screw is there and stops when the screw is taken away, we do not say that the screw is the equivalent of the machine. For correspondence to be equivalence, it would be necessary that to any part of the machine a definite part of the screw should correspond—as in a literal translation in which each chapter renders a chapter, each sentence a sentence, each word a word. Now, the relation of the brain to consciousness seems to be entirely different. Not only does the hypothesis of an equivalence between the psychical state and the cerebral state imply a downright absurdity, as we have tried to prove in a former essay,[1] but the facts, examined without prejudice, certainly seem to indicate that the relation of the psychical to the physical is just that of the machine to the screw. To speak of an equivalence between the two is simply to curtail, and make almost unintelligible, the Spinozistic or Leibnizian metaphysic. It is to accept this philosophy, such as it is, on the side of Extension, but to mutilate it on the side of Thought. With Spinoza, with Leibniz, we suppose the unifying synthesis of the phenomena of matter achieved, and everything in matter explained mechanically. But, for the conscious facts, we no longer push the synthesis to the end. We stop half-way. We suppose consciousness to be coextensive with a certain part of nature and not with all of it. We are thus led, sometimes to an "epiphenomenalism" that associates consciousness with certain particular vibrations and puts it here and there in the world in a sporadic state, and sometimes to a "monism" that scatters consciousness into as

[1] "Le Paralogisme psycho-physiologique" (*Revue de métaphysique et de morale*, Nov. 1904, pp. 895-908). Cf. *Matière et mémoire*, Paris, 1896, chap. i.

many tiny grains as there are atoms; but, in either case it is to an incomplete Spinozism or to an incomplete Leibnizianism that we come back. Between this conception of nature and Cartesianism we find, moreover, intermediate historical stages. The medical philosophers of the eighteenth century, with their cramped Cartesianism, have had a great part in the genesis of the "epiphenomenalism" and "monism" of the present day.

These doctrines are thus found to fall short of the Kantian criticism. Certainly, the philosophy of Kant is also imbued with the belief in a science single and complete, embracing the whole of the real. Indeed, looked at from one aspect, it is only a continuation of the metaphysics of the moderns and a transposition of the ancient metaphysics. Spinoza and Leibniz had, following Aristotle, hypostatized in God the unity of knowledge. The Kantian criticism, on one side at least, consists in asking whether the whole of this hypothesis is necessary to modern science as it was to ancient science, or if part of the hypothesis is not sufficient. For the ancients, science applied to *concepts*, that is to say, to kinds of *things*. In compressing all concepts into one, they therefore necessarily arrived at a *being*, which we may call Thought, but which was rather thought-object than thought-subject. When Aristotle defined God the νοησεως νοήσις, it is probably on νοήσεως, and not on νόησις that he put the emphasis. God was the synthesis of all concepts, the idea of ideas. But modern science turns on laws, that is, on relations. Now, a relation is a bond established by a mind between two or more terms. A relation is nothing outside of the intellect that relates. The universe, therefore, can only be a system of laws if phenomena have passed beforehand through the filter of an intellect. Of course, this intellect might be that of a being infinitely superior to man, who would found the materiality of things at the same time that he bound them together: such was the hypothesis of Leibniz and of Spinoza. But it is not necessary to go so far, and, for the effect we have here to obtain, the human intellect is enough: such is precisely the Kantian solution. Between the dogmatism of a Spinoza or a Leibniz and the criticism of Kant there is just the same distance as between "it may be maintained that—" and "it suffices that—." Kant stops this dogmatism on the incline that was making it slip too far toward the Greek metaphysics; he reduces to the strict minimum the hypothesis which is necessary in order to suppose the physics of Galileo indefinitely extensible. True, when he speaks of the human intellect, he means neither yours nor mine: the unity of nature comes indeed from the human understanding that unifies, but the unifying function that operates here is impersonal. It imparts itself to our individual consciousnesses, but it transcends them. It is much less than a substantial God; it is, however, a little more than the isolated work of a man or even than the collective work of humanity. It does not exactly lie within man; rather, man lies within it, as in an atmosphere of intellectuality which his consciousness breathes. It is, if we will, a *formal* God, something that in Kant is not yet divine, but which tends to become so. It became so, indeed, with Fichte. With Kant, however, its principal rôle was to give to the whole of our science a relative and *human* character, although of a humanity already somewhat deified. From this point of view, the criticism of Kant consisted chiefly in limiting the dogmatism of his predecessors, accepting their conception of science and reducing to a minimum the metaphysic it implied.

But it is otherwise with the Kantian distinction between the matter of knowledge and its form. By regarding intelligence as pre-eminently a faculty of establishing relations, Kant attributed an extra-intellectual origin to the terms between which the relations are established. He affirmed, against his immediate predecessors, that knowledge is not entirely resolvable into terms of intelligence. He brought back into philosophy—while modifying it and carrying it on to another plane—that essential element of the philosophy of Descartes which had been abandoned by the Cartesians.

Thereby he prepared the way for a new philosophy, which might have established itself in the extra-intellectual matter of knowledge by a higher effort of intuition. Coinciding with this matter, adopting the same rhythm and the same movement, might not consciousness, by two efforts of opposite direction, raising itself and lowering itself by turns, become able to grasp from within, and no longer perceive only from without, the two forms of reality, body and mind? Would not this twofold effort make us, as far as that is possible, re-live the absolute? Moreover, as, in the course of this operation, we should see intellect spring up of itself, cut itself out in the whole of mind, intellectual knowledge would then appear as it is, limited, but not relative.

Such was the direction that Kantianism might have pointed out to a revivified Cartesianism. But in this direction Kant himself did not go.

He *would* not, because, while assigning to knowledge an extra-intellectual matter, he believed this matter to be either coextensive with intellect or less extensive than intellect. Therefore he could not dream of cutting out intellect in it, nor, consequently, of tracing the genesis of the understanding and its categories. The molds of the understanding and the understanding itself had to be accepted as they are, already made. Between the matter presented to our intellect and this intellect itself there was no relationship. The agreement between the two was due to the fact that intellect imposed its form on matter. So that not only was it necessary to posit the intellectual form of knowledge as a kind of absolute and give up the quest of its genesis, but the very matter of this knowledge seemed too ground down by the intellect for us to be able to hope to get it back in its original purity. It was not the "thing-in-itself," it was only the refraction of it through our atmosphere.

If now we inquire why Kant did not believe that the matter of our knowledge extends beyond its form, this is what we find. The criticism of our knowledge of nature that was instituted by Kant consisted in ascertaining what our mind must be and what Nature must be *if* the claims of our science are justified; but of these claims themselves Kant has not made the criticism. I mean that he took for granted the idea of a science that is one, capable of binding with the same force all the parts of what is given, and of co-ordinating them into a system presenting on all sides an equal solidity. He did not consider, in his *Critique of Pure Reason*, that science became less and less objective, more and more symbolical, to the extent that it went from the physical to the vital, from the vital to the psychical. Experience does not move, to his view, in two different and perhaps opposite ways, the one conformable to the direction of the intellect, the other contrary to it. There is, for him, only *one* experience, and the intellect covers its whole ground. This is what Kant expresses by saying that all our intuitions are sensuous, or, in other words, infra-intellectual. And this would have to be admitted, indeed, if our science presented in all its parts an equal objectivity. But suppose, on the contrary, that science is less and less objective, more and more symbolical, as it goes from the physical to the psychical, passing through the vital: then, as it is indeed necessary to perceive a thing somehow in order to symbolize it, there would be an intuition of the psychical, and more generally of the vital, which the intellect would transpose and translate, no doubt, but which would none the less transcend the intellect. There would be, in other words, a supra-intellectual intuition. If this intuition exists, a taking possession of the spirit by itself is possible, and no longer only a knowledge that is external and phenomenal. What is more, if we have an intuition of this kind (I mean an ultra-intellectual intuition) then sensuous intuition is likely to be in continuity with it through certain intermediaries, as the infra-red is continuous with the ultra-violet. Sensuous intuition itself, therefore, is promoted. It will no longer attain only the phantom of an unattainable thing-in-itself. It is (provided we bring to it certain indispensable corrections) into the absolute itself that it will introduce us. So long as it was regarded as the only material of our science, it reflected back on all science something of the relativity which strikes a scientific knowledge of spirit; and thus the perception of bodies, which is the beginning of the science of bodies, seemed itself to be relative. Relative, therefore, seemed to be sensuous intuition. But this is not the case if distinctions are made between the different sciences, and if the scientific knowledge of the spiritual (and also, consequently, of the vital) be regarded as the more or less artificial extension of a certain manner of knowing which, applied to bodies, is not at all symbolical. Let us go further: if there are thus two intuitions of different order (the second being obtained by a reversal of the direction of the first), and if it is toward the second that the intellect naturally inclines, there is no essential difference between the intellect and this intuition itself. The barriers between the matter of sensible knowledge and its form are lowered, as also between the "pure forms" of sensibility and the categories of the understanding. The matter and form of intellectual knowledge (restricted to its own object) are seen to be engendering each other by a reciprocal adaptation, intellect modeling itself on corporeity, and corporeity on intellect.

But this duality of intuition Kant neither would nor could admit. It would have been necessary, in order to admit it, to regard duration as the very stuff of reality, and consequently to distinguish between the substantial duration of things and time spread out in space. It would have been necessary to regard space itself, and the geometry which is immanent in space, as an ideal limit in the direction of which material things develop, but which they do not actually attain. Nothing could be more contrary to the letter, and perhaps also to the spirit, of the *Critique of Pure Reason*. No doubt, knowledge is presented to us in it as an ever-open roll, experience as a push of facts that is forever going on. But, according to Kant, these facts are spread out on one plane as fast as they arise; they are external to each other and external to the mind. Of a knowledge from within, that could grasp them in their springing forth instead of taking them already sprung, that would dig beneath space and spatialized time, there is never any question. Yet it is indeed beneath this plane that our consciousness places us; there flows true duration.

In this respect, also, Kant is very near his predecessors. Between the non-temporal, and the time that is spread out in distinct moments, he admits no mean. And as there is indeed no intuition that carries us into the non-temporal, all intuition is thus found to be sensuous, by definition. But between physical existence, which is spread out in space, and non-temporal existence, which can only be a conceptual and logical existence like that of which metaphysical dogmatism speaks, is there not room for consciousness and for life? There is, unquestionably. We perceive it when we place ourselves in duration in order to go from that duration to moments, instead of starting from moments in order to bind them again and to construct duration.

Yet it was to a non-temporal intuition that the immediate successors of Kant turned, in order to escape from the Kantian relativism. Certainly, the ideas of becoming, of progress, of evolution, seem to occupy a large place in their philosophy. But does duration really play a part in it? Real duration is that in which each form flows out of previous forms, while adding to them something new, and is explained by them as much as it explains them; but to deduce this form directly from one complete Being which it is supposed to manifest, is to return to Spinozism. It is, like Leibniz and Spinoza, to deny to duration all efficient action. The post-Kantian philosophy, severe as it may have been on the mechanistic theories, accepts from mechanism the idea of a science that is one and the same for all kinds of reality. And it is nearer to mechanism than it imagines; for though, in the consideration of matter, of life and of thought, it replaces the successive degrees of complexity, that mechanism supposed by degrees of the realization of an Idea or by degrees of the objectification of a Will, it still speaks of degrees, and these degrees are those of a scale which Being traverses in a single direction. In short, it makes out the same articulations in nature that mechanism does. Of mechanism it retains the whole design; it merely gives it a different coloring. But it is the design itself, or at least one half of the design, that needs to be remade.

If we are to do that, we must give up the method of *construction*, which was that of Kant's successors. We must appeal to experience—an experience purified, or, in other words, released, where necessary, from the molds that our intellect has formed in the degree and proportion of the progress of our action on things. An experience of this kind is not a non-temporal experience. It only seeks, beyond the spatialized time in which we believe we see continual rearrangements between the parts, that concrete duration in which a radical recasting of the whole is always going on. It follows the real in all its sinuosities. It does not lead us, like the method of construction, to higher and higher generalities—piled-up stories of a magnificent building. But then it leaves no play between the explanations it suggests and the objects it has to explain. It is the detail of the real, and no longer only the whole in a lump, that it claims to illumine.

That the thought of the nineteenth century called for a philosophy of this kind, rescued from the arbitrary, capable of coming down to the detail of particular facts, is unquestionable. Unquestionably, also, it felt that this philosophy ought to establish itself in what we call concrete duration. The advent of the moral sciences, the progress of psychology, the growing importance of embryology among the biological sciences—all this was bound to suggest the idea of a reality which *endures* inwardly, which is duration itself. So, when a philosopher arose who announced a doctrine of evolution, in which the progress of matter toward perceptibility would be traced together with the advance of the mind toward rationality, in which the complication of correspondences between the external and the internal would be followed step by step, in which change would become the very substance of things—to him all eyes were turned. The powerful attraction that Spencerian evolutionism has exercised on contemporary thought is due to that very cause. However far Spencer may seem to be from Kant, however ignorant, indeed, he may have been of Kantianism, he felt, nevertheless, at his first contact with the biological sciences, the direction in which philosophy could continue to advance without laying itself open to the Kantian criticism.

But he had no sooner started to follow the path than he turned off short. He had promised to retrace a genesis, and, lo! he was doing something entirely different. His doctrine bore indeed the name of evolutionism; it claimed to remount and redescend the course of the universal becoming; but, in fact, it dealt neither with becoming nor with evolution.

We need not enter here into a profound examination of this philosophy. Let us say merely that *the usual device of the Spencerian method consists in reconstructing evolution with fragments of the evolved.* If I paste a picture on a card and then cut up the card into bits, I can reproduce the picture by rightly grouping again the small pieces. And a child who working thus with the pieces of a puzzle-picture, and putting together unformed fragments of the picture finally obtains a pretty colored design, no doubt imagines that he has *produced* design and color. Yet the act of drawing and painting has nothing to do with that of putting together the fragments of a picture already drawn and already pâinted. So, by combining together the most simple results of evolution, you may imitate well or ill the most complex effects; but of neither the simple nor the complex will you have retraced the genesis, and the addition of evolved to evolved will bear no resemblance whatever to the movement of evolution.

Such, however, is Spencer's illusion. He takes reality in its present form; he breaks it to pieces, he scatters it in fragments which he throws to the winds; then he "integrates" these fragments and "dissipates their movement." Having *imitated* the Whole by a work of mosaic, he imagines he has retraced the design of it, and made the genesis.

Is it matter that is in question? The diffused elements which he integrates into visible and tangible bodies have all the air of being the very particles of the simple bodies, which he first supposes disseminated throughout space. They are, at any rate, "material points," and consequently unvarying points, veritable little solids: as if solidity, being what is nearest and handiest to us, could be found at the very origin of materiality! The more physics progresses, the more it shows the impossibility of representing the properties of ether or of electricity—the probable base of all bodies—on the model of the properties of the matter which we perceive. But philosophy goes back further even than the ether, a mere schematic figure of the relations between phenomena apprehended by our senses. It knows indeed that what is visible and tangible in things represents our possible action on them. It is not by dividing the evolved that we shall reach the principle of that which evolves. It is not by recomposing the evolved with itself that we shall reproduce the evolution of which it is the term.

Is it the question of mind? By compounding the reflex with the reflex, Spencer thinks he generates instinct and rational volition one after the other. He fails to see that the specialized reflex, being a terminal point of evolution just as much as perfect will, cannot be supposed at the start. That the first of the two terms should have reached its final form before the other is probable enough; but both the one and the other are *deposits* of the evolution movement, and the evolution movement itself can no more be expressed as a function solely of the first than solely of the second. We must begin by mixing the reflex and the voluntary. We must then go in quest of the fluid reality which has been precipitated in this twofold form, and which probably shares in both without being either.

At the lowest degree of the animal scale, in living beings that are but an undifferentiated protoplasmic mass, the reaction to stimulus does not yet call into play one definite mechanism, as in the reflex; it has not yet choice among several definite mechanisms, as in the voluntary act; it is, then, neither voluntary nor reflex, though it heralds both. We experience in ourselves something of this true original activity when we perform semi-voluntary and semi-automatic movements to escape a pressing danger. And yet this is but a very imperfect imitation of the primitive character, for we are concerned here with a mixture of two activities already formed, already localized in a brain and in a spinal cord, whereas the original activity was a simple thing, which became diversified through the very construction of mechanisms like those of the spinal cord and brain. But to all this Spencer shuts his eyes, because it is of the essence of his method to recompose the consolidated with the consolidated, instead of going back to the gradual process of consolidation, which is evolution itself.

Is it, finally, the question of the correspondence between mind and matter? Spencer is right in defining the intellect by this correspondence. He is right in regarding it as the end of an evolution. But when he comes to retrace this evolution, again he integrates the evolved with the evolved—failing to see that he is thus taking useless trouble, and that in positing the slightest fragment of the actually evolved he posits the whole—so that it is vain for him, then, to pretend to make the genesis of it.

For, according to him, the phenomena that succeed each other in nature project into the human mind images which represent them. To the relations between phenomena, therefore, correspond symmetrically relations between the ideas. And the most general laws of nature, in which the relations between phenomena are condensed, are thus found to have engendered the directing principles of thought, into which the relations between ideas have been integrated. Nature, therefore, is reflected in mind. The intimate structure of our thought corresponds, piece by piece, to the very skeleton of things—I admit it willingly; but, in order that the human mind may be able to represent relations between phenomena, there must first be phenomena, that is to say, distinct facts, cut out in the continuity of becoming. And once we posit this particular mode of cutting up such as we perceive it today, we posit also the intellect such as it is today, for it is by relation to it, and to it alone, that reality is cut up in this manner. Is it probable that mammals and insects notice the same aspects of nature, trace in it the same divisions, articulate the whole in the same way? And yet the insect, so far as intelligent, has already something of our intellect. Each being cuts up the material world according to the lines that its action must follow: it is these lines of *possible action* that, by intercrossing, mark out the net of experience of which each mesh is a fact. No doubt, a town is composed exclusively of houses, and the streets of the town are only the intervals between the houses: so, we may say that nature contains only facts, and that, the facts once posited, the relations are simply the lines running between the facts. But, in a town, it is the gradual portioning of the ground into lots that has determined at once the place of the houses, their general shape, and the direction of the streets: to this portioning we must go back if we wish to understand the particular mode of subdivision that causes each house to be where it is, each street to run as it does. Now, the cardinal error of Spencer is to take experience already allotted as given, whereas the true problem is to know how the allotment was worked. I agree that the laws of thought are only the integration of relations between facts. But, when I posit the facts with the shape they have for me today, I suppose my faculties of perception and intellection such as they are in me today; for it is they that portion the real into lots, they that cut the facts out in the whole of reality. Therefore, instead of saying that the relations between facts have generated the laws of thought, I can as well claim that it is the form of thought that has determined the shape of the facts perceived, and consequently their relations among themselves: the two ways of expressing oneself are equivalent; they say at bottom the same thing. With the second, it is true, we give up speaking of evolution. But, with the first, we only speak of it, we do not think of it any the more. For a true evolutionism would propose to discover by what *modus vivendi*, gradually obtained, the intellect has adopted its plan of structure, and matter its mode of subdivision. This structure and this subdivision work into each other; they are mutually complementary; they must have progressed one with the other. And, whether we posit the present structure of mind or the present subdivision of matter, in either case we remain in the evolved: we are told nothing of what evolves, nothing of evolution.

And yet it is this evolution that we must discover. Already, in the field of physics itself, the scientists who are pushing the study of their science furthest incline to believe that we cannot reason about the parts as we reason about the whole; that the same principles are not applicable to the origin and to the end of a progress; that neither creation nor annihilation, for instance, is inadmissible when we are concerned with the constituent corpuscles of the atom. Thereby they tend to place themselves in the concrete duration, in which alone there is true generation and not only a composition of parts. It is true that the creation and annihilation of which they speak concern the movement or the energy, and not the imponderable medium through which the energy and the movement are supposed to circulate. But what can remain of matter when you take away everything that determines it, that is to say, just energy and movement themselves? The philosopher must go further than the scientist. Making a clean sweep of everything that is only an imaginative symbol, he will see the material world melt back into a simple flux, a continuity of flowing, a becoming. And he will thus be prepared to discover real duration there where it is still more useful to find it, in the realm of life and of consciousness. For, so far as inert matter is concerned, we may neglect the flowing without committing a serious error: matter, we have said, is weighted with geometry; and matter, the reality which *descends*, endures only by its connection with that which *ascends*. But life and consciousness are this very ascension. When once we have grasped them in their essence by adopting their movement, we understand how the rest of reality is derived from them. Evolution appears and, within this evolution, the progressive determination of materiality and intellectuality by the gradual consolidation of the one and of the other. But, then, it is within the evolutionary movement that we place ourselves, in order to follow it to its present results, instead of recomposing these results artificially with fragments of themselves. Such seems to us to be the true function of philosophy. So understood, philosophy is not only the turning of the mind homeward, the coincidence of human consciousness with the living principle whence it emanates, a contact with the creative effort: it is the study of becoming in general, it is true evolutionism and consequently the true continuation of science—provided that we understand by this word a set of truths either experienced or demonstrated, and not a certain new scholasticism that has grown up during the latter half of the nineteenth century around the physics of Galileo, as the old scholasticism grew up around Aristotle.

INDEX

(Compiled by the Translator)